THE FRACKING DEBATE

CENTER ON GLOBAL ENERGY POLICY SERIES

CENTER ON GLOBAL ENERGY POLICY SERIES

JASON BORDOFF, SERIES EDITOR

Making smart energy policy choices requires approaching energy as a complex and multifaceted system in which decision makers must balance economic, security, and environmental priorities. Too often, the public debate is dominated by platitudes and polarization. Columbia University's Center on Global Energy Policy at SIPA seeks to enrich the quality of energy dialogue and policy by providing an independent and nonpartisan platform for timely analysis and recommendations to address today's most pressing energy challenges. The Center on Global Energy Policy Series extends that mission by offering readers accessible, policy-relevant books that have as their foundation the academic rigor of one of the world's great research universities.

Robert McNally, *Crude Volatility: The History and the Future of Boom-Bust Oil Prices* (2017)

Richard Nephew, *The Art of Sanctions: A View from the Field* (2018)

DANIEL RAIMI

THE FRACKING DEBATE

The Risks, Benefits, and Uncertainties of the Shale Revolution

COLUMBIA UNIVERSITY PRESS
NEW YORK

Columbia University Press
Publishers Since 1893
New York Chichester, West Sussex
cup.columbia.edu

Library of Congress Cataloging-in-Publication Data
Names: Raimi, Daniel, author.
Title: The fracking debate : the risks, benefits, and uncertainties of the shale revolution / Daniel Raimi.
Description: New York : Columbia University Press, [2017] | Series: Center On Global Energy Policy series | Includes bibliographical references and index.
Identifiers: LCCN 2017024122 (print) | LCCN 2017026192 (ebook) | ISBN 9780231545716 (e-book) | ISBN 9780231184861 (cloth : alk. paper)
Subjects: LCSH: Hydraulic fracturing—Environmental aspects—Risk assessment. | Hydraulic fracturing—Social aspects.
Classification: LCC TD195.G3 (ebook) | LCC TD195.G3 R34 2017 (print) | DDC 363.11/96223383—dc23
LC record available at https://lccn.loc.gov/2017024122

Columbia University Press books are printed on permanent and durable acid-free paper.

Printed in the United States of America

Cover design: Jordan Wannemacher
Cover photo: Mark Thiessen/© GettyImages

For my parents, Jane and Fred

CONTENTS

ACKNOWLEDGMENTS

I am grateful for the support of many friends, colleagues, and family members who have contributed directly or indirectly to making this book a reality. Thanks to my parents, Jane and Fred, and my sister Lisa for a lifetime of love; to Kari Barsness, Robin Smith, Mary Penny Kelley, and Trina Ozer for bringing me into the DENR fold and giving me an introduction to shale gas; to Melissa Levine, Leena Lalwani, and Charles Watkinson at the University of Michigan for their help understanding how publishing works; to Jeff Collett at Colorado State University, Rick Neitzel at the University of Michigan, and Donna Vorhees at the Health Effects Institute for discussions on air emissions and potential health effects; to Mike Teague, the Oklahoma secretary of energy and environment, for comments on seismicity in Oklahoma; to Eric Kort from the University of Michigan and Mark Brownstein from the Environmental Defense Fund for discussions of methane, ethane, and other air emissions; to Rick McCurdy for graciously allowing me to tell his family's story about Alpine High; to Mary Barber, Gail Dreyfuss, and Max Raimi for their thoughts on an early version of this book; and to Tom Murphy at Penn State University for comments on a near-finished draft. Thank you to the institutions I've been lucky enough to call home over the past several years: the Duke University Energy Initiative, the University of Michigan Energy Institute, the University of Michigan Gerald R. Ford School of Public Policy, and Resources for the Future. Thank you to Evan Michelson and the Alfred P. Sloan Foundation for supporting related research into oil and gas development, which gave me the opportunity to see the oil and gas fields of the United States. Thank you to Barry Rabe at the University of Michigan for planting the seed of this book, even if he doesn't remember doing so. Thank you to Michael Levi for suggesting I send my idea for this book to Columbia University, and thanks to Jason Bordoff at Columbia's Center on Global Energy Policy for helping make it a reality. More than anyone else, Richard Newell at Resources for the Future has given me the opportunity to learn about energy and the environment. He has also been a wonderful teacher, fantastic collaborator, and friend. Jeremy Boak from the Oklahoma Geological Survey provided detailed comments on an early draft of this book and continued to provide constructive feedback during the revision process. I thank him for devoting so much time and for his help making this a better book. Bridget Flannery-McCoy at Columbia University Press has been a fantastic editor, making sure that my arguments are structured logically and helping refine my explanations of issues that can be confusing or counterintuitive. Most importantly, thank you to my wife, Kaitlin Raimi, who has learned more about the oil and gas industry than she ever wanted to know. Her love, friendship, and support have made me happier than I ever thought I could be.

THE FRACKING DEBATE

1

INTRODUCTION

I wasn't expecting a polka band.

It was February 2016, and I was in Houston, Texas, for CERAWeek, the annual energy-industry confab that brought together highfliers from around the world: heads of state, corporate icons, academic experts, and others came to discuss the state of the oil and gas industry, learn about new technologies, and assess what might lie ahead. With prices at crushing lows—a barrel of crude oil was trading for about thirty dollars, down from one hundred dollars two years earlier—I'd expected a somber air. But at the week's opening evening, a grand cocktail party with the crème de la crème of the industry, sponsored by the German corporate giant Siemens, it was as though I had entered an alternate reality, one in which the industry was thriving, crude oil and natural gas prices were near all-time highs, and celebration was the order of the day.

At least that's what it felt like as the six-person polka band started up, the tuba belting out lumps of bass notes as servers sliced thick, heavy slabs of beef at the two prime-rib carving stations. Hotel staff stood behind black-draped bars, pouring top-shelf liquor, fine wines, and microbrews. At other tables, three-tiered silver platters gleamed atop white tablecloths, boasting rich arrays of German-themed hors d'oeuvres. Regardless of the challenges the industry faced, tonight was a night to have fun, to network, and, inevitably, to commiserate about the pain of low prices.

Had it not been for the crushing prices, there would have been every reason for domestic oil and gas companies to celebrate: production of both oil and natural gas was near all-time highs, and the United States had reclaimed the title of world's largest producer. Just ten years earlier, most had expected the country's decades-long decline in production to continue indefinitely and, with it, an ever-deepening reliance on energy imported from other continents. But things had changed.

As I'll describe over the course of this book, the U.S. oil and gas industry has, over the past decade or so, combined a suite of technological breakthroughs with incremental improvements that have pushed production to levels beyond what even the most optimistic forecaster would have dreamed. These innovations brought the industry to new heights, took it to new corners of the United States, and sparked controversies no one had anticipated.

It was this revolution, the "shale revolution," that had brought me to the Hilton Americas in Houston. CERAWeek was something of a culmination for me: a chance to meet some of the industry's high rollers after spending years getting to know the small-timers of the oilfield and traveling the many back roads of the shale revolution.

IN THE OILFIELD

My journey started in the summer of 2011. I was a graduate student at Duke University in Durham, North Carolina, and was spending the summer interning at the state's Department of Environment and Natural Resources (which has since been reorganized and is now known as the Department of Environmental Quality). While I was there, the legislature asked the agency to write a report on the potential for shale gas development in the state. I volunteered for the job and, despite my lack of experience with the oil and gas industry, was tasked with writing a portion of the report.

As I began my research, two themes quickly emerged: one detailing a list of horror stories about the dangers of fracking and a second narrative about the glories of the American energy renaissance. As I learned more and began talking with friends about the myths and realities of fracking, I started to field an array of questions. At dinner parties, out having drinks with friends, and even over my first Christmas dinner with my soon-to-be mother- and father-in-law, I heard variations on the same few themes: Were fracking chemicals safe? Was fracking causing earthquakes? Was fracking contaminating water across the country?

Everyone who learned I was doing research on fracking had questions, and most of these questions rested on the presumption that the process was inherently dangerous. Many assumed that fracking was wreaking havoc on landscapes and communities across the United States. Gingerly, I would tiptoe through the history, risks, and unknowns of shale development. At the end of the conversation, I would occasionally get a final question: Could I recommend any books that offered an accessible discussion of the full issue? I didn't have a good answer.

My experience that summer piqued my interest in oil and gas, but in North Carolina, there was no opportunity to witness shale development firsthand. Luckily, I began working on a research project a couple years later that provided exactly this opportunity. The project focused on local government finances: how increased (or decreased) oil and gas development had affected local revenues and services and how those issues played into state tax policy. The position took me to all the major shale gas and oil "hot spots": western North Dakota, northeastern Pennsylvania, southern Texas, western Texas, Louisiana, Colorado, Wyoming, and more. I spent most of my days driving between small towns, stopping wherever I could snap a nice photo of a drilling rig, pump jack, or flare stack. In each town, I would spend an hour or two interviewing local government officials and eating at local restaurants, learning about people's experience with shale development.

Once the interviews were through and I had finished my work for the day, I had a full evening on my hands to explore. If I wasn't too tired after a long drive or day of meetings, I'd find a place to have dinner and sit at the bar, listening to the local chatter. Often, I'd join the conversation, or—just as often—my neighbor or the bartender would ask, "What brings you to town?" But they knew what I was going to say: it was the oilfield. The only question remaining was whether I was an engineer, a driller, a geologist, an exec, or something else.

When I told them I was doing research, I got a variety of reactions. Some were skeptical, assuming that I was in town to look for environmental damage and try to put them out of business. Others, after learning I was researching government issues, let loose on the evergreen topics of regulation, taxes, and the unwelcome hand of the federal government. But after a few minutes of conversation, most of my neighbors at the bar were happy to talk about what it's been like to live or work in the oilfield through the boom and, as the heady days of 2013 turned to wary 2014 and eventually to the doldrums of 2015, through the bust.

I talked with hundreds of locals and out-of-towners, pipeliners and frackers, drill-baby-drillers and no-fracking-wayers about the oilfield. During those conversations, I learned as much about how fracking has affected the United States as I did from the hundreds of peer-reviewed articles, books, and news reports that I read during the same time.

A few years into my work, I was in Ann Arbor, Michigan, giving a talk at the Ford School of Public Policy (where I would eventually become a lecturer) describing a recent paper on tax policy related to oil and gas development. I flew in the night before and was having dinner with Barry Rabe, the professor who had invited me to town. As we ate and talked, he asked about the places where I'd traveled to do my research. When I described the kinds of conversations I'd been having—both my interviews during the day and my informal conversations with locals at night—Barry asked if I'd thought about writing a book. I agreed it'd be a great way to document the experience but didn't think I had the time. I was busy finishing my research, looking forward to taking on new projects, and hesitant to wade into the caustic terrain that characterizes much of the debate over fracking.

But as the months went on and I traveled to even more oil and gas regions—Oklahoma, New Mexico, Ohio, California, Alaska, and others—I gathered more stories and met more people. The map in figure 1.1 shows each of the major oil- and gas-producing regions of the United States that I have visited over the past several years. This is the oilfield. The dots on the map indicate each of the more than 200,000 oil and gas wells in the United States that have been drilled directionally or horizontally, rather than vertically. These wells, particularly the horizontal ones, are where the shale revolution has taken place, and most of them have been fracked. I'll describe what fracking is, and what it isn't, in chapter 2, along with a discussion of why horizontal drilling, in particular, is an important part of the story.

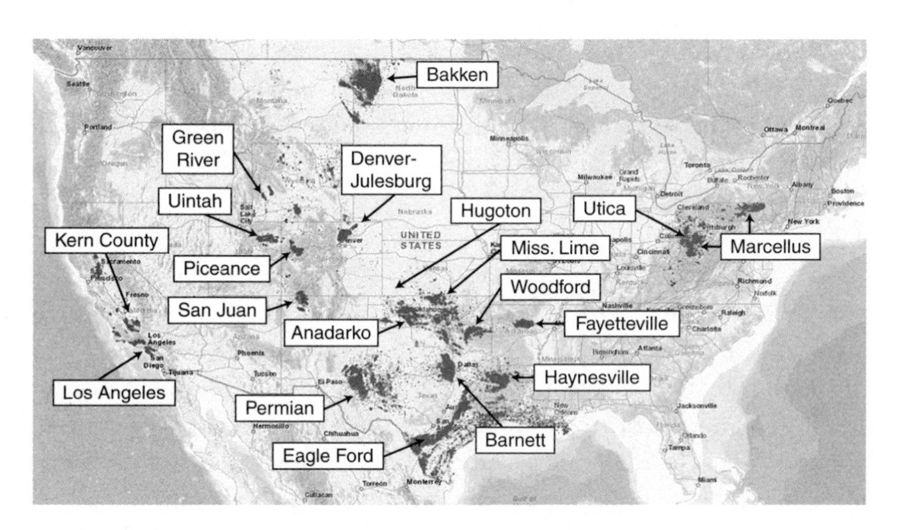

FIGURE 1.1 An annotated map of the author's travels from 2013 through 2015

Map indicates all horizontally and directionally drilled wells. Data not available for Alaska. Annotations by the author.

Source: Drilling Info database.

A frequent mistake of some commentators is to assume that every place where oil and gas are pulled out of the ground looks the same. That couldn't be further from the truth. Oil and gas production happens in and around some of the biggest U.S. cities, such as Los Angeles and Fort Worth, where wells pop up in backyards, along freeways, or next to fast-food drive-throughs. Oil and gas drilling happens in some of the most rural parts of the country, such as western North Dakota and southwestern Wyoming, where cows and tumbleweeds easily outnumber people. It happens in the fast-growing suburbs north of Denver and on the outskirts of Pittsburgh, where drilling rigs cast shadows over new apartment buildings and strip malls encroach on land previously devoted to farming, ranching, and oil and gas.

To understand how the oil and gas industry fits into the lives of people in these diverse regions, it's helpful to know at least a little bit about them. The stories that begin each chapter of this book will take us to a few of these places, but if you've never visited an oilfield personally, think about adding the Permian basin, Utica shale, or another producing region to your list of travel destinations. A few days of driving through oil and gas country, coupled with conversations across a barstool, can teach you as much as dozens of journal articles and research reports.

THE CHALLENGE OF UNCERTAINTY

The goal of this book is to present a full view of shale development in the United States, drawing both from the ground-level experiences I've had and from the academic literature that continues to develop. In my opinion, a full view of shale development must recognize three critical facts: first, the shale revolution has created benefits; second, the shale revolution has caused damage and imposed costs; and, finally, there are still a number of important uncertainties. In this book, I will try to give readers an honest perspective on the scale and importance of all three elements.

One of the most challenging things about writing a book about fracking (or any other timely policy issue) is that many of the topics and questions are moving targets. While research on some issues is well developed and the lessons fairly straightforward, a number of unanswered questions remain, and uncertainty runs like a steady stream through some of the most important elements of this book.

For some readers, the existence of these uncertainties may elicit the question: With all we don't know about fracking, why take the risk? This question leads to a discussion of the *precautionary principle*, the notion that until all the risks of an activity are fully understood, it's best to wait.

One does not have to agree with this philosophy to understand that it is a coherent way of looking at the world. Nations such as France and Germany, along with the state of New York, have essentially adopted this approach, banning fracking or imposing extended moratoriums. Compared with their U.S. counterparts, European policy makers tend to embrace the precautionary principle more readily when considering new technologies. A prime example is genetically modified foods, or GMOs. European nations have adopted far stricter regulations than the United States for growing and selling these crops.

But if you apply the precautionary principle to GMOs or to oil and gas development, should you apply it to everything?

Consider mobile phones. Some fear that the widespread use of these devices may increase the likelihood of developing cancer. The National Cancer Institute at the National Institutes of Health stated in 2016 on its website: "Studies have thus far not shown a consistent link between cell phone use and cancers of the brain, nerves, or other tissues of the head and neck. More research is needed because cell phone technology and how people use cell phones have been changing rapidly."[1] For an individual or policy maker prone to the precautionary principle, this statement might justify a ban on cell phones. At the very least, it provides a pretty compelling reason for the risk-averse to avoid them. What's more, there are a number of areas where we know that cell phones have caused damage. For example, research and common experience have shown that the proliferation of mobile phones has increased vehicle and pedestrian accidents[2] and reduced the quality of life for anyone within earshot of an inconsiderate cell phone user (but I digress).

My analogy may be a little flippant, but the principle is clear: fracking produces benefits, risks, and uncertainties. But that's nothing new. Policy makers and individuals balance benefits and risks every day. Cell phones provide enormous benefits to virtually every person in the United States and billions more around the world. Fracking, which serves a completely different purpose in the economy, probably provides smaller economic benefits than mobile technology, but, as I will describe in chapter 9, the economic effects have been enormous, affecting energy consumers around the world.

In short, the precautionary approach may be appropriate in some cases, but it can't be used to justify many of the trade-offs that you and I make every day. If the precautionary principle is the lodestar by which we navigate risks and rewards, it quickly becomes difficult to justify using any number of new technologies.

MY GOAL

While I will identify and examine these uncertainties, there are also benefits and risks that we understand pretty well. In each chapter, I'll assess these three elements one question at a time. Those questions are the ones that, in my experience, come up most often when I talk with colleagues, students, friends, and relatives in classrooms, across dinner tables, and in bars. Along the way, I'll try to give a flavor of the places I've traveled and provide a sense of some of the hundreds of people I've met in the oilfield.

While no one person can be an expert on all of the questions addressed in this book, I've been lucky enough to work closely over the past several years with some of the leading experts in the fields of environmental and energy economics, hydrogeology, environmental law, and state regulation. I've also gotten to know some of the leading figures in academia, industry, and advocacy. Conversations with many of these experts have helped shape my understanding of each issue, and my hope is that each chapter hits the most important elements of the fracking debate.

Still, I'm certain that neither side of this often-heated debate will agree with every aspect of this book. In fact, I am confident that *both* sides will take issue with certain portions. But that probably means I'm striking a reasonable balance. As I'll try to demonstrate, neither side has a monopoly on the right answers. Where oil and gas advocates make misleading or demonstrably false arguments, I'll try to put them into context and, where necessary, correct them. Where antifracking advocates make unsubstantiated claims or stoke unwarranted fear, I'll do the same.

Debate surrounding each of the topics described in this book will—and should—continue. Barring some major and unforeseen shift in energy policy, economics, or politics, shale development and fracking is going to continue for the foreseeable future, and a well-informed public debate should be a part of the policy-making process. But it needs to be *well informed*. After the election of 2016, when basic facts (let alone nuanced interpretations of complex questions) were sometimes scorned, I understand this may be a Sisyphean task.

Indeed, as I've watched and listened to the often vitriolic and misleading debates over fracking, there tends to be far more heat than light. Well-crafted studies that do not align with one side's preexisting views are rejected out of hand, and any shred of evidence confirming that same preexisting view is trumpeted as "*the* Science." The average person doesn't know what to think or whom to trust.

The truth is that there are real risks, real benefits, and real uncertainties surrounding fracking. My goal is for readers of this book to come away with a better understanding of all three.

2

WHAT IS FRACKING?

For most of the year, Beaumont, Texas, is hot and humid, sprawling long and flat across the terrain of southeastern Texas. Located along the Neches River, about thirty-five miles north of the Sabine Pass, Beaumont grew up alongside a natural shipping channel that opens into the Gulf of Mexico. Beaumont, along with its southern neighbor Port Arthur, hosts some of the largest oil refineries in the world. These behemoths, which operate twenty-four hours a day, process billions of barrels of oil each year, heating, separating, storing, and shipping some of the roughly twenty million barrels of petroleum products (gasoline, diesel, jet fuel, heating oil, etc.) that Americans consume every day.

I had a free day between meetings in Texas and needed to get from the Haynesville shale region in northwestern Louisiana to the Eagle Ford (pronounced as one word: EE-gull-ferd) shale play in South Texas. (An oil or gas "play" is simply a geographical region where substantial deposits of oil and gas are found. A "play" can refer to a single geologic formation or multiple formations within the same region.) Beaumont wasn't exactly on the way, but it wasn't a huge detour, either. And I had a reason for visiting: Spindletop.

The name Spindletop holds a central place in American energy history. In 1901, it was there, in Beaumont, where a group of prospectors brought into production the first giant oilfield in the United States. Spindletop refers to the hill upon which these developers made their discovery. Beneath the hill, trapped under a salt dome roughly 1,000 feet below the surface, were massive stores of oil.

While oil had been produced in the United States since the mid-1800s, Spindletop signaled the beginning of a new and more prolific era in American oil. At a time when many wells produced only ten to twenty barrels per day (one barrel equals forty-two gallons), the first successful well drilled at Spindletop initially produced as much as 100,000 barrels a day. Almost overnight, the state of Texas became the epicenter of the national energy market, soon growing even further to become the pumping heart of global oil production.

Beaumont became a boomtown, polluted and overcrowded, filled with bars and brothels. The scenes around Spindletop Hill were chaotic and fetid, with hustlers and hucksters rubbing elbows, all looking to strike it rich. Howard Hughes, who made his fortune by inventing a drilling bit that cut through rock more efficiently than its predecessors (he would eventually be portrayed by Leonardo DiCaprio in the film *The Aviator*), described the cacophony of the oil business during the early days of Spindletop:

> Beaumont in those days was no place for a divinity student. The reek of oil was everywhere. It filled the air, it painted the houses, it choked the lungs and stained men's souls. Such another excitement will not be seen for a generation. It will take that length of time to get together an equal number of fools and "come-on's" at one spot. I turned greaser and sank into the thick of it. Rough neck, owner, disowner, promoter, capitalist and "mark"—with each I can claim kin, for I have stood in the steps of each.[1]

The early wells at Spindletop produced oil at astonishing rates—these are the "gushers" we often imagine when we think of oil wells. This popular perception of how oil is produced has been reinforced by films such as *Giant* or *There Will Be Blood*, where a single well produces thousands of barrels of oil per day—"blowing gold all over the place," in the words of Daniel Plainview, the fictional oilman portrayed by Daniel Day-Lewis in Paul Thomas Anderson's masterpiece.

But most oil wells are not like Spindletop. Most don't blow gold all over the place. Indeed, even Spindletop didn't gush for long, and today, most of the oil available under the hill at Spindletop has been drained,

FIGURE 2.1 **Spindletop's Hogg Swayne section, 1902**

Source: Texas Energy Museum. Used with permission.

and the hill has deflated to a flat, fallow field marked by a silver pole denoting the site of the original gusher. The first commercial oil wells, drilled in northeastern Pennsylvania in the mid-1800s, weren't gushers either. Instead they produced modest quantities of ten to twenty barrels per day. In these plays, the natural pressures in the underground rock formations where oil was held were mild, and the black stuff flowed slowly. Because of those low pressures, each well tapped just the small area around the bottom of the drilled hole, and production was modest.

But as Russell Gold describes in his book *The Boom*,[2] a Civil War veteran named Edward A. L. Roberts adapted his knowledge of explosives to enhance greatly the production of oil wells in Pennsylvania. Just six years after the first commercial well drilled by "Colonel" Edwin Drake began producing in 1859, Roberts demonstrated that exploding ordinance at the bottom of an existing well could break apart the rocks that were stubbornly hoarding oil and allow more liquid to travel into and up the well. This technology was hugely successful, and the Roberts Petroleum Torpedo Company went on to make Mr. Roberts a very rich man.

This technique, dubbed "shooting" the well, was not without danger on the surface. Indeed, mishandling Roberts's torpedoes could—and did—lead to explosions, injury, and death for some working in the early days of the oilfield.[3]

Despite these risks, oil producers since the industry's inception have looked for ways to coax ever more oil from the earth. The gusher at

Spindletop is a prime example of an oil well that relied solely on natural pressure, but other wells, whether drilled in the 1860s or the 2010s, require additional coaxing. And while the early explosives deployed by Roberts's company are a far cry from today's hydraulic fracturing, the historical lesson is simple: blowing stuff up underground is nothing new for the oil industry.

THE ADVENT OF MODERN FRACKING

Since the founding of the Roberts Petroleum Torpedo Company, generations of engineers and investors have sought new ways to produce more oil and gas from a single well. Many of these experiments, generally labeled "well stimulation," have involved underground explosives. Fracking, short for "hydraulic fracturing," is one such technique. I'll detail the fracking process later in this section, but for now, let's define fracking as the process of injecting a fluid deep underground to create cracks in a rock formation, which helps more oil and gas flow into a well.

In 1949, the oilfield-service firm Halliburton, contracted by Stanolind Oil, was the first to fracture a well hydraulically, meaning that instead of simply lowering an explosive into the earth, they would pump fluids at high pressure, hoping to create small cracks in the rock and increase the flow of—in this case—natural gas. Along with reducing the risks associated with handling nitroglycerin at the well site, hydraulic fracturing had other advantages. The process created cracks in the rock that extended over a greater distance than earlier explosive-based well-stimulation efforts. These longer cracks allowed oil and gas to flow into the well from further afield, increasing overall production and profitability.[4] While the operation did not employ conventional explosives, it nonetheless involved some seriously flammable fluids: a mixture of napalm and gasoline pumped roughly 2,500 feet underground into the massive Hugoton gas field in western Kansas.[5]

Since that time, hydraulic fracturing has been applied to millions of wells around the United States and the world, and these "frack jobs," as they are often called,[6] vary substantially from place to place. Depending on the type of rock formation where the oil and gas are located, different mixtures of water, chemicals, and sand are used to create cracks in the rocks and allow oil and gas to flow more easily.

While the term "fracking" is often associated with natural gas, in recent years more oil wells have been fracked in the United States than natural gas wells. This association in the public's mind between fracking and natural gas is partly because the first major shale plays developed using hydraulic fracturing primarily produced natural gas. However, the notion that fracking is relevant only for natural gas is incorrect. When we're talking about fracking, we're talking about both gas *and* oil production.

So if fracking has been around for more than sixty years, why did it suddenly become controversial in the 2000s? The answer is simple: hydraulic fracturing, coupled with other more recent technological advances, enabled companies to develop oil and gas from *shale* profitably.[7]

Shale formations are the ovens where oil and gas are created, as decaying organic material—typically microbes and plant matter from ancient lakes and oceans—becomes buried under layer after layer of new sediment. Over tens of millions of years, these decaying organisms are compressed as they are buried and are subject to increasing temperatures as they are pushed deeper into the earth's crust.

In many cases, the shale formations have—again over many millions of years—expelled some of this oil and gas from the "oven" up toward the surface through natural pathways in the layers of rock. In most cases, the oil and gas become trapped during ascent by another rock layer that they cannot pass through (in some cases, the oil and gas make it all the way up and surface through "seeps," exemplified by the famous LaBrea tar pits in Los Angeles or by Jed Clampett's fictional "bubbling crude"). These underground traps, like the salt dome at Spindletop, are the formations that have been tapped for most of the industry's history, and production from these reservoirs is today termed "conventional."

The industry had long known that source rocks were the keepers of vast stores of oil and gas. However, these rocks were in most cases too "tight" to tap economically.[8] Unlike conventional oil and gas formations such as limestone or sandstone, shale and other "tight" rocks are impermeable; that is, oil and gas do not easily flow through them. In a conventional (more permeable) reservoir, fluids and gases can move easily from one part of the rock to another, like water moving through a sponge. When a well is drilled into one of these formations, the existing underground pressure pushes oil and gas toward the lower-pressure area created by the well. These underground pressures then push oil, gas, and water to the surface (water is almost always found alongside oil and gas underground). If the pressures are great enough, you might get a gusher like the one seen at Spindletop. But for a well drilled into a shale formation—even a shale formation that holds large quantities of oil and/or gas—the impermeability of the rock prevents large volumes of those gases and liquids from moving naturally toward the well.

Despite the technical challenges involved in tapping these impermeable formations, shale remained attractive to some within the industry. The vast potential for oil and gas production from shale and other tight rocks—if engineers could just find a way to coax it out—led to continued research and experimentation. Starting in the 1960s, the U.S. government funded multiple research programs related to shale and tight-gas production, perhaps most strikingly in Project Gasbuggy, Project Rulison, and Project Rio Blanco, three experiments that employed nuclear devices to release gas from tight formations in the late 1960s and early 1970s in remote areas of New Mexico and Colorado.[9] These experiments were in some ways similar to the early "shooting" of wells in Pennsylvania, albeit at a much larger scale. While the technique of producing natural gas using nuclear detonations did not proliferate, other federal-government initiatives, tax credits, and price premiums continued to encourage companies to search for ways to develop "unconventional" sources of oil and natural gas.[10]

Mitchell Energy, a natural gas producer in the Dallas–Fort Worth region (though a minnow compared to behemoths such as Exxon or Chevron), was one such interested company. Its founder, George Mitchell, encouraged his engineers to keep experimenting with shale despite years of failure and frustration. As described in *The Frackers*, Gregory Zuckerman's history of early shale developers,[11] Mitchell needed new supplies of gas to fulfill existing contracts and was convinced that shale could provide the needed volumes. Although many in his own company thought him foolhardy, Mitchell continued pressing his engineers to find a way to make shale profitable.

The company had for years relied on a gel-based fluid to create fractures in the Barnett shale, a formation that lay 5,000 to 8,000 feet beneath Fort Worth and the surrounding area. But the results were poor. Cracks in the shale opened up by the gel closed up quickly, inhibiting the flow of gas and making most of these wells economic losers. Mitchell's engineers tweaked the composition of the fluid they used at each new well, watching carefully as the results of their experiments rolled in.

But the most consequential change was the result of an accident. While fracking one Barnett well, an equipment malfunction created a fracturing fluid that was primarily water based and therefore much "slicker" than the goopy, gel-based fluids. Prevailing wisdom held that water-based fluids wouldn't work in shale formations, as the clay embedded in the rock would absorb water, then swell to plug any fractures. But the company's engineers didn't shut down the frack job. Instead, they let it play out, in part because water was much cheaper than the chemical-heavy gels they had been using.

As it turned out, the slickwater frack produced encouraging results. And while some in the company continued to believe it was a crazy idea, Mitchell began using this water-based fracking fluid in all of its Barnett shale wells. Experimenting with hundreds of wells, they adjusted the amount of sand in the mixture, pumped it at greater pressures, and used greater volumes of water.

Over time, as the engineers continued to tweak, the wells got better and better, producing more and more gas. In time, Mitchell Energy was consistently making a profit from its Barnett wells, and other companies started to catch on. In 2001, Mitchell was bought for $3.1 billion by Devon Energy, a larger company focused on natural gas. The shale revolution had begun.

HOW FRACKING AND SHALE DEVELOPMENT AS A WHOLE WORK

Along with Mitchell's new approach to fracking, other technological advances have been crucial in enabling the widespread extraction of natural gas and oil from shale. Perhaps the most important of these advances, which include high-resolution seismic imaging (to understand better the location and types of formations below the surface) and more aggressive financial engineering (which allowed some early shale players to lease vast tracts of land prior to developing the play),[12] was the perfection of directional and horizontal drilling.

Since the industry's inception, most oil and gas wells have been drilled straight down—vertically—aimed at an oil or gas deposit directly below the surface. But as technology improved, some wells came to be drilled at an angle, a process known as directional drilling or, colloquially, "slant drilling." This type of drilling might be necessary if an oil or gas field lay beneath a lake or a city, where it would be too difficult to deploy a drilling rig.

Over decades, the precision of directional drilling has improved dramatically. Dozens of articles from petroleum-engineering journals detail the development of the technology, and Devon Energy, which acquired Mitchell, had been known as an industry leader in directional and—more importantly—horizontal drilling,[13] the practice of drilling vertically to a certain depth, then slowly turning the well to run parallel to the surface.

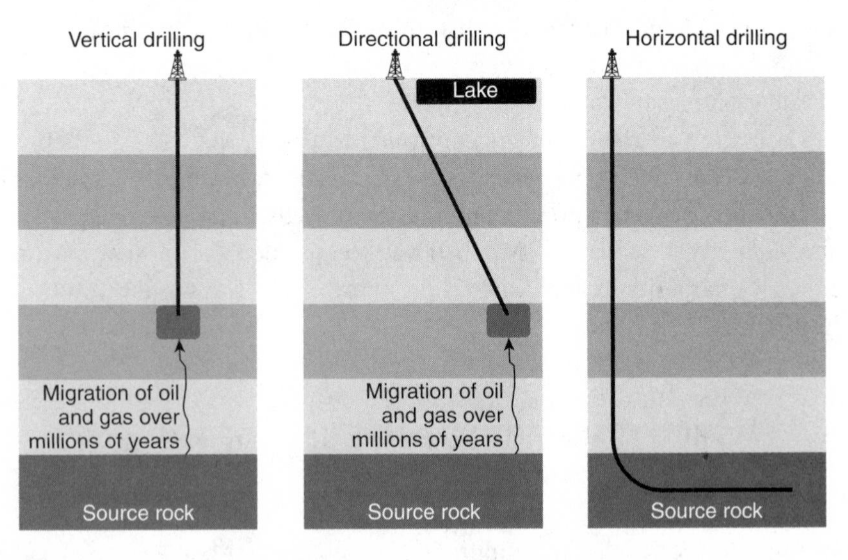

FIGURE 2.2 **Vertical, directional, and horizontal drilling**

Source: Author.

Horizontal drilling has been essential to the success of shale producers for a simple reason: permeability. Because of the impermeability of shale, the oil and gas won't naturally flow to the well, as it might in a conventional formation. Instead, the well needs to go to *it*.

The combination of horizontal drilling and hydraulic fracturing, along with the other advances already mentioned, would prove to be the keys that unlocked widespread shale development. To illustrate the entire process, I'll take a hypothetical company—Rollins Energy—and describe how it might go about finding and developing a new shale play.

ROLLINS ENERGY

Rollins Energy, along with the rest of the oil and gas industry, had known for decades that there was a large shale formation called "the Bridge" about 6,000 feet below the surface. Based on rock samples taken by geologists years ago, Rollins Energy estimated how much oil and gas was contained within the rocks and believed that they could profitably develop the Bridge.

So Rollins hired ten "landmen," professional men or women who scour land records at local courthouses to determine who owns which tracts. Once the landmen identify a certain number of adjoining properties, they set out to visit those landowners to persuade them to lease their land. If they succeed and a lease is signed, landowners are paid a "bonus" of anywhere from $100 to $30,000 per acre of leased land,[14] giving Rollins the right to drill on their land within five years (or some other negotiated time frame). If Rollins does not drill within the specified number of years, the lease expires, and the landowner keeps his or her bonus. Rollins also negotiates a royalty rate with the property owner, ranging from 12.5 to 25 percent[15] of the value of the oil and gas produced from their property (12.5 percent is the minimum royalty rate in most—if not all—states).[16]

Once Rollins acquires enough acres to drill a few wells, it selects a site and builds a well pad. Heavy construction machines such as bulldozers and graders clear an area on one of the leased properties, and crews move in to construct a three-to-five-acre well pad, typically consisting of compacted gravel and sand. A drilling rig then trundles into town and is hoisted atop the pad, followed by a dozen or so support vehicles and workers to operate the rig.

Over the course of a week or more (this timing can vary depending on factors such as the depth of the formation and the length of the horizontal section of the well), the rig operates day and night, drilling down about 5,000 feet before it reaches the "kickoff point," where it begins to bend toward the Bridge shale layer, about 6,000 feet down (shale formations can be found at any depth, but most of the production from shale today comes from depths between 3,000 and 10,000 feet). By 6,000 feet, the drill bit is moving horizontally. This horizontal portion of the well is known as the "lateral," and it bores about a mile through the shale. Over time, companies have drilled longer and longer laterals, sometimes stretching as far as two miles or more. After the well itself is drilled, steel pipe known as "casing" is cemented into place to carry the oil and gas from the Bridge to the surface. Once the drilling phase is complete and the casing is laid, the drilling rig trundles out of town and toward its next job.

At the well site, Rollins prepares to "complete" the well. First, a device known as a perforating gun, or "perf gun," is lowered into the well. Throughout the length of the one-mile lateral, the perf gun detonates small charges, blasting holes in the casing and in the neighboring shale. These holes in the steel will soon be the conduits through which oil and gas will flow from the Bridge and into the well.

Next, it's time for fracking. Over the course of the previous days, dozens of trucks or, in some cases, pipelines have delivered millions of gallons of fresh or recycled water, which has been stored at the well site. Other trucks have delivered chemicals, and still other trucks have delivered many tons of sand.

These three key ingredients, water, sand, and chemicals, are mixed together and pumped by large diesel-powered engines into the well at extremely high pressure. The mixture fills the well, then begins to flow out of the well and into the holes created by the perf gun. As pressure slowly increases from the heavy-duty pumps at the surface, the fracking fluid flows into the shale and creates fractures that open up the Bridge, extending the small holes created by the perf gun and creating new networks of cracks throughout the rock. These fractures are extremely narrow but range in length from 100 to 1,000 feet (in rare cases extending as far as 2,000 feet).[17] These narrow pathways overcome the problem of impermeability by creating networks for oil or gas to flow from the shale into the well.[18]

Much of the sand pumped into the well remains in those pathways, acting as a "proppant" by keeping them open and allowing oil and gas to continue flowing after pressure from the fluids is reduced. Quartz sand with a high proportion of silica is typically used because of its strength and ability to withstand the intense heat and pressure far below the surface. In some cases, more high-tech ceramic beads act as proppants in place of or alongside this sand.

Once Rollins has finished fracking the Bridge, natural underground pressure forces some of the fluids to flow up the well and back to the surface. The large volumes of water that flow back to the surface mingle with chemicals from the fracking fluid as well as with compounds from the Bridge, including brine (salty water), oils, and—in some places—naturally occurring radioactive materials, or NORMs. Rollins captures this "flowback," piping it from the well and into a storage tank located on the well pad. (In some places, flowback is stored in open pits lined with plastic or other materials to prevent the liquids from seeping into the ground. As I'll describe in the next chapter, these protections are not always successful.)

As the company puts the finishing touches on the well—affixing surface equipment and building pipelines to move the oil or gas, it enters its life of production: oil and gas will flow out of the Bridge, up the well, and into pipelines that take it to waiting customers. Along with oil and gas, the Bridge produces a substantial amount of salty water, which continues to be piped into storage tanks on site.

This completes the most active phase of well development, and although the entire process is often referred to as fracking, the full cycle of developing an oil or gas well involves much more than just the hydraulic-fracturing stage. This process of developing a well—from when the pad is constructed to when the oil and gas are sent to buyers—can last anywhere from one to several months.

The well will produce oil and gas for decades, sometimes for forty years or longer, depending on how productive it is. A very productive shale well might produce 400,000 barrels of oil and 1.2 billion cubic feet of natural gas in its first year,[19] an output worth anywhere from around $15 million to $45 million depending on oil and natural gas prices (remember, shale wells can produce oil, natural gas, or both). The well is at its most prolific in the initial weeks and months following its completion. Production declines sharply over the first six months or so, then declines more slowly over time, producing at a slower and steadier rate for most of its life.

As with "conventional" oil and gas formations, companies drilling into shale formations will not be able to extract 100 percent of the oil and gas trapped in the rock. The portion they do recover is called the "recovery rate." These recovery rates vary from formation to formation, ranging from roughly 2 percent to 20 percent in some locations.[20] For more permeable conventional plays, where oil and gas flow more easily into a well, recovery rates are typically higher.

From time to time, workers in large trucks come by the well site to empty the storage tanks holding oil or brine. Smaller trucks also appear regularly to make sure everything is working properly, with workers emerging to check valves, tanks, or other equipment. For the most part, the well sits quietly for ten to thirty years, pumping out some combination of oil and gas, which in most cases is moved through a pipeline (in places without a dense pipeline network, oil is usually moved by truck) to processing facilities.

From there, the oil and gas will be made into the products we use every day. Oil refineries will turn crude oil into gasoline, diesel, or jet fuel, and some oil will be used to make plastics. Natural gas–processing facilities will separate the raw gas into its component parts, the most prevalent of which is methane, which is delivered through pipelines to homes, businesses, and power plants to provide heating, generate electricity, and more.

THE DIFFERENCE BETWEEN FRACKING AND "FRACKING"

While the oil and gas industry has developed a deep understanding of fracking over its decades of use, the concept is still being defined in the public imagination, and both pro- and antifracking advocates have sought to define the word to suit their purposes. As a result, it can sometimes be confusing, even for experts, to determine what someone means when they use the word "fracking."

Many of the chapters in this book will show that some of the most common fears associated with fracking are actually more likely to be caused by some other part of the process: casing and cementing a well, moving oil through pipelines, disposing of wastewater, and so on. To be clear, fracking refers to hydraulic fracturing, which is just one of the many steps required to produce oil and gas from shale and other tight sources. However, the term "fracking" has often been used to refer to a much broader set of activities, and opponents of the fossil-fuel industry writ large often refer to the "fracking industry" instead of the "oil and gas industry," "fracking wells" instead of "oil and gas wells," "fracking pipelines" instead of "natural gas pipelines," and so on.

Making this strategy explicit, Josh Fox, an antifracking filmmaker (who will appear again in the following chapter), argued at a 2017 public hearing in Pennsylvania that "we have to make sure that our definition of fracking does not stop at the gas wells. Our definition of fracking has to include the power plants and the pipelines . . . because we are under siege as a nation right now from the fossil-fuel industry."[21] If "fracking" refers to the entire set of activities related to oil and gas development, it becomes easier for advocates to argue that "fracking" is the cause of any high-profile case of pollution, whether or not those cases are the result of hydraulic fracturing itself. Pipeline rupture? Blame fracking. Local groundwater pollution from surface spills? Blame fracking.

And if the public believes that "fracking"—rather than some more prosaic issue such as a pipeline leak or an oil spill at the surface—causes the damage to human health or the environment, it becomes easier to rally support behind a movement to ban fracking (meaning the discrete process of hydraulic fracturing).

For example, a 2016 headline from *DeSmog*, a website focused on climate and environmental issues, touted a court ruling with the headline "Dimock Water Contamination Verdict Prompts Calls for Federal Action on Fracking."[22] This despite the fact that the water pollution in question was caused by errors in the casing and cementing of the well, a risk for any oil or gas well, be it "conventional" or fracked. Another headline from the left-leaning website *ThinkProgress* describes a study that found the same problem (of errors in well casing and cementing) with the headline "Study Links Water Contamination to Fracking Operations in Texas and Pennsylvania."[23]

In the next chapter, I'll describe how a variety of phases in the oil- and gas-production process can cause pollution, but I'll also describe how hydraulic fracturing has not been the leading culprit.

On the profracking side of the story, industry advocates are typically more precise with their terminology when discussing the environmental risks of fracking. *Energy in Depth*, an industry-backed advocacy group, typically employs a narrow definition of "fracking" when discussing the risks of oil and gas development.

For example, a 2015 study examining a sample of sixty-four water wells in Pennsylvania found traces of fracking fluids in local groundwater, most likely as a result of spills by oil and gas companies on the surface.[24] *Energy in Depth* trumpeted this finding with the headline "New Study Finds Fracking Has Not Contaminated Drinking Water,"[25] while another industry-advocacy website covered the study with a story titled "Fracking Does Not Contaminate Drinking Water."[26]

Technically, these headlines are accurate because the researchers found that spills and leaks at the surface, not the underground process of hydraulic fracturing, were the likely causes of water pollution. This narrow definition of fracking enables industry supporters such as Senator James Inhofe to state that "my state of Oklahoma has led the way on hydraulic fracturing regulations, and just like the rest of the nation, we have yet to see an instance of ground water contamination."[27] However, an audience that understands "fracking" to mean all phases of oil and gas production will not understand this distinction and is thus less likely to be concerned about the environmental risks caused by drilling and other industry activities.

Complicating the matter further, when industry supporters want to tout the *benefits* of oil and gas production (such as lower energy prices or other economic benefits), their definition of fracking begins to balloon. For example, numerous articles from *Energy in Depth* employ an expansive definition of fracking when describing the economic benefits of the shale revolution, with headlines such as "Leading Business Group Says Fracking Provides 'Substantial Economic Benefits' for Colorado" and "Energy Secretary Ernest Moniz: Fracking Is Good for the Economy and Environment."[28]

As a result, readers of *DeSmog* or *ThinkProgress* end up with one definition of fracking, while readers of *Energy in Depth* come away with another. The word "fracking" comes to resemble the famous Rubin's vase optical illusion, where a single black-and-white image looks like a vase to one person, but another sees two faces in profile. Like a Rorschach test, different audiences see different things. The confused language surrounding fracking is a major reason for the deep distrust in which each side of the debate holds the other. Advocates on each side hear "fracking" and picture two completely different sets of activities, depending on the goals of their argument. And both sides use whichever term best suits their agenda, ensuring continued distrust between warring factions and continued confusion among the public.

WHAT FRACKING HAS DONE FOR U.S. OIL AND GAS PRODUCTION

As this book will describe, there are a lot of uncertainties surrounding the risks and benefits of oil and gas development, including environmental and health effects, economic impacts, the effects on climate change, and a variety of other important issues. But at least one thing is certain: shale development, enabled by fracking and other technological improvements, has revolutionized oil and gas production in the United States.

For most of the twentieth century, the United States was the world's leading producer and consumer of natural gas, followed by Russia. But as the 1970s and 1980s dawned and U.S. production failed to keep up with demand, many in the business, academic, and policy community became concerned that the United States would soon need to import large quantities of gas from other countries. Although the United States was still producing a lot of natural gas, these domestic sources were steadily declining, and imports (mostly via pipelines from Canada) were steadily growing. With this trend expected to continue, investors poured billions of dollars into terminals along the East and Gulf coasts designed to receive shipments of liquefied natural gas from suppliers such as Qatar, Russia, and Trinidad and Tobago.

In the early and mid-2000s, as the demand from power plants and other users grew, domestic supply remained relatively flat. The result was a surge in natural gas prices, which helped push up the costs of home heating, electricity, and many other energy-related products (see figure 2.3). Around the same time, oil prices began a steady march upward, amplifying concerns that high energy prices would slow economic growth in the United States for the foreseeable future.

Because of the relatively small size of Mitchell Energy and the other companies beginning to extract large quantities of gas from shale, most

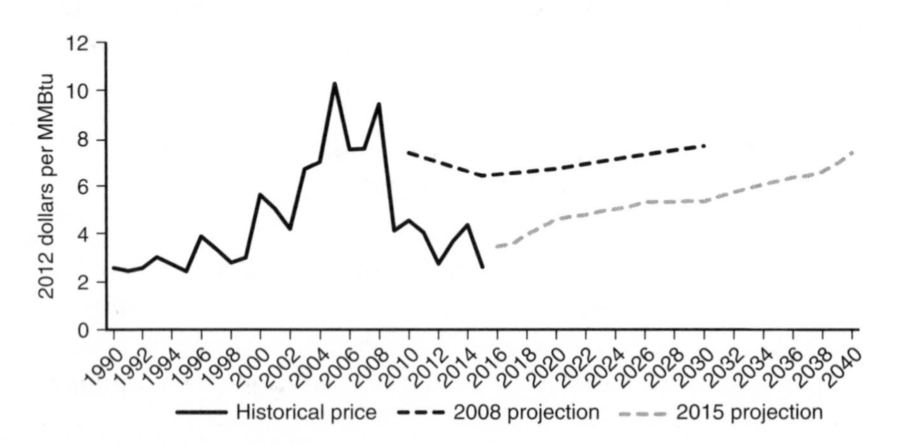

FIGURE 2.3 U.S. natural gas prices and projections

Sources: U.S. EIA, Annual Energy Outlook, Reference Case (2015), https://www.eia.gov/outlooks/aeo/pdf/0383(2015).pdf; U.S. EIA, Annual Energy Outlook, Reference Case (2008), http://alternativeenergy.procon.org/sourcefiles/annual-energy-outlook-2008.pdf; U.S. EIA, "Henry Hub Natural Gas Spot Price," (2015), http://www.eia.gov/dnav/ng/hist/rngwhhdm.htm.

experts—even many within the oil and gas industry—didn't realize that the shale revolution had already begun. Others thought that Mitchell Energy's success in the Barnett was a one-off: that it couldn't be replicated elsewhere.

But by the mid-2000s, production from the Barnett was substantial, and other companies, led by Chesapeake Energy, began to lease millions of acres of land for shale development.[29] Over the next decade, natural gas production from shale grew at an unprecedented rate.

Drilling rigs and fracking crews began to move out from the Barnett and into other regions of the South and Southwest: the Woodford shale in southeastern Oklahoma, the Fayetteville shale in northern Arkansas, and the Haynesville shale, which straddles the border between northwestern Louisiana and northeastern Texas. By the early 2010s, massive new natural gas plays were coming into production in other parts of the country, led by the Marcellus shale in Pennsylvania and West Virginia, the Eagle Ford shale in southern Texas, and the Utica shale in eastern Ohio.

In 2000, natural gas production from shale and other tight formations (such as the Green River basin in southwestern Wyoming and the Piceance basin in western Colorado) was a little over 1 trillion cubic feet per year, or about 5 percent of the total U.S. production. By 2005, it had roughly doubled to 2 trillion cubic feet and 11 percent of total production. But by 2014, these sources together had grown to more than 13 trillion cubic feet per year, accounting for half of the U.S. total. As figure 2.4 shows, overall production in the United States easily surpassed all-time highs, resulting in a rapid decline in natural gas prices. Looking forward, natural gas prices are projected to stay relatively low for decades to come.

But natural gas has not been the only fuel affected by fracking and the development of shale resources. The same types of horizontal drilling and hydraulic-fracturing techniques that enabled large-scale production of shale gas has also reinvigorated U.S. oil production in a manner few believed possible.

Thanks to the nineteenth-century oil discoveries in Pennsylvania, Texas, and elsewhere, the United States led the world in crude production for roughly a century.[30] But like that of natural gas, domestic oil production plateaued and began to decline in the 1970s and 1980s, convincing

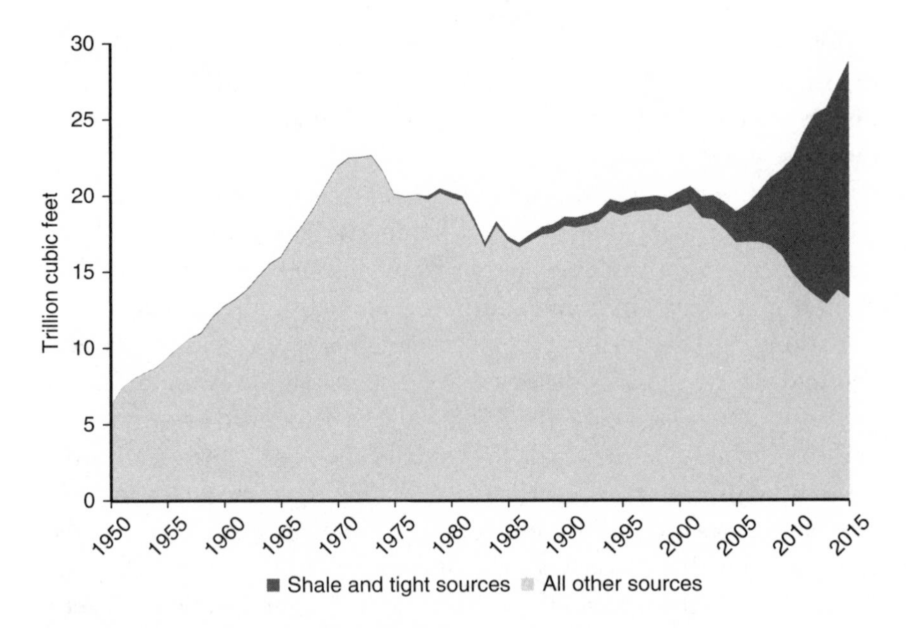

FIGURE 2.4 **U.S. natural gas production**

Drilling information for shale and "tight" sources include production from the following basins and formations: Green River basin, Piceance basin, Barnett shale, Woodford shale, Fayetteville shale, Haynesville shale, Marcellus shale, Eagle Ford shale, Utica-Point Pleasant shale.

Sources: U.S. EIA, "Natural Gas Marketed Production" (2015), https://www.eia.gov/naturalgas/; Drilling Info (DI Desktop): Reserve reports by formation and basin (2016).

many that the United States would become ever more reliant on imports for the remainder of the oil age. This was not the first time that fears of "peak oil" had taken hold of the public's imagination,[31] but the slow and steady decline in production had convinced most experts that America's glory days of oil production had passed.

However, shale development changed all of that in rapid fashion. From a low point of 5 million barrels per day in 2008, domestic crude oil production has grown by 70 percent, to 9.4 million barrels in 2015. While a variety of formations have contributed to this increase, production has been led by three regions: the Eagle Ford shale in southern Texas, the Bakken and Three Forks formations in western North Dakota and

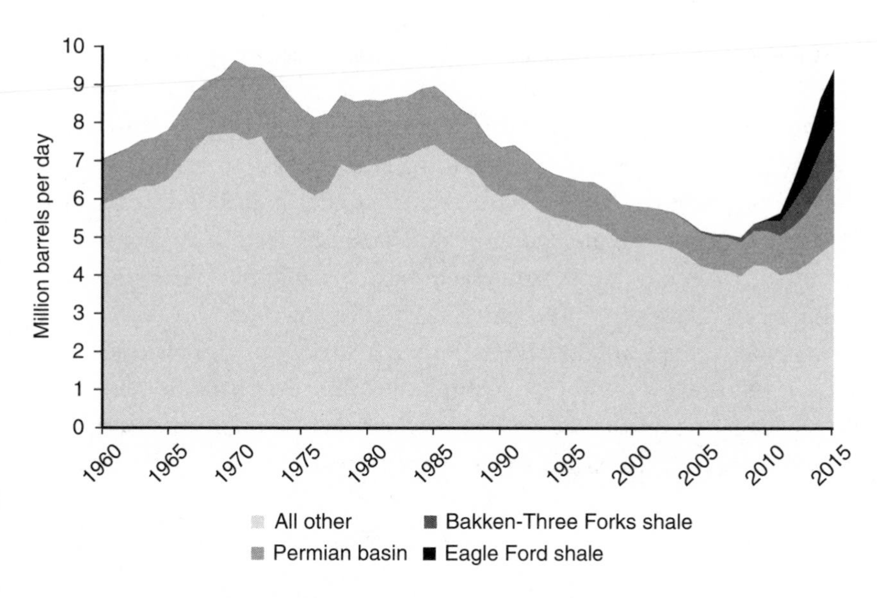

FIGURE 2.5 **U.S. oil production**

Sources: U.S. EIA, "Field Production of Crude Oil" (2015), https://www.eia.gov/petroleum/; Drilling Info (DI Desktop): Reserve reports by formation and basin (2016).

eastern Montana, and the Permian basin in western Texas and eastern New Mexico. The Permian had long been a leader in U.S. production, but most oil over the decades had come from vertical wells tapping conventional reservoirs. The surge in oil production depicted in figure 2.5 came about through unconventional development of the various shale and other tight formations that lay beneath or between those conventional plays.

LOOKING BACK, LOOKING FORWARD

Back at Spindletop, I wandered through a dusty courtyard and between clapboard buildings, snapping photos of early-twentieth-century oilfield equipment. Rusted and ragged, the drilling rigs of the day were powered by steam engines, which drove a motor to operate the drill bit. These contraptions were held together by wooden beams and hemp ropes and featured ten-foot-tall wooden wheels. I wondered what the drillers at Spindletop would have thought of the technologies of today's oilfield.

Some things have changed a lot: computerized control rooms, high-volume fracturing fleets, precise drilling equipment that can hit a target the size of a football from two miles away. Other issues have remained: rowdy boomtowns sprung from the prairie, underground detonations of various sizes, and fortunes made seemingly overnight.

As we examine the major issues related to fracking and oil and gas development more broadly, there will be a variety of complexities, nuances, and questions without easy answers. But there is no denying that the shale revolution has had an unprecedented effect on oil and gas in the United States. The implications of this revolution, however, are far from straightforward.

SUMMING UP

Virtually since its inception, the oil and gas industry has been blowing stuff up underground. Just a few years after the first oil wells came into production in Pennsylvania in the mid-1800s, the industry began detonating various explosives in attempts to coax more oil out of the ground and into their wells. Today's hydraulic-fracturing, or fracking, operations are a larger-scale, more precise, and more water-intensive adaptation of these activities. In recent years, this newer, bigger, and more sophisticated version of "stimulating" oil- and gas-bearing rock formations has been combined with precise horizontal drilling, better underground-mapping techniques, and other advances to increase oil and gas production dramatically in the United States.

3

DOES FRACKING CONTAMINATE WATER?

Dimock Township, a rural community tucked into the wooded hills of northeastern Pennsylvania, is the most notorious place on the fracking map. It's been featured in numerous documentaries, combed over by news crews, and invoked frequently by antifracking advocates. The reason for all this attention is a high-profile case of water contamination associated with natural gas drilling in 2008, during the early days of the development of the Marcellus shale. In no small part because of Dimock, water contamination has perhaps been the leading fear related to fracking and the shale revolution. For antifracking advocates, Dimock has become much more: a symbol of the environmental damage fracking has wrought.

Fears have been stoked largely by a single viral image: the flaming faucet. Josh Fox's Academy Award–nominated 2010 film, *Gasland*, features numerous scenes from along Carter Road, a narrow dirt lane in Dimock. Residents hold jars of murky water from their private wells, which they argue were polluted by fracking. Fox states that the residents had been able to light their water on fire, though they can't perform the feat on the day of the film crew's visit.

Several minutes later, the scene shifts to Weld County, Colorado, where we get our first glimpse of the flaming faucet. A middle-aged man in a white T-shirt, gaunt and with sandy gray hair, holds a lighter up to his flowing tap. After a couple of seconds . . . whoosh! The water ignites, engulfing the sink in a fireball, singing eyebrows, and raising more than a few questions.

Because of the film's steady focus on chemicals involved in the fracking process, viewers are led to believe that these chemicals have migrated into water supplies in Dimock and Weld County, making daily activities such as showering, drinking a glass of water, or washing the dishes a hazard and a health threat.

As I approached Dimock, driving north through Susquehanna County and replaying in my head the scenes from *Gasland*, I half expected to see something out of Fritz Lang's *Metropolis*: industrialism run amok against a pastoral Pennsylvania backdrop. But as I drove into town, the first yard sign that I saw read: "Leave us alone! Our water is fine!" Clearly this house had been visited more than once by researchers or reporters. Carter Road, however, was a different story. Nearly every yard sported antifracking messages such as "We can't drink money!" and "Don't frack with our water!"

What was going on in Dimock? Had fracking contaminated the water or not? As it turns out, the simple narratives advocated by partisans on either side of the debate (Fracking pollutes water! Fracking doesn't pollute water!) were misleading. The truth was far more complex.

UP FROM THE DEPTHS?

Perhaps the greatest fear the average citizen has regarding fracking is the concern that chemicals injected underground will contaminate water supplies. Throughout much of the antifracking community, the implication of the exploding sinks featured in *Gasland* is that fracking chemicals inevitably infiltrate aquifers, turning clean drinking water into flammable poison.

To back up this point, antifracking artists often forgo a sense of scale when illustrating the well-drilling and fracking process, simplifying and distorting the events taking place underground. Many depictions of fracking show an oil or gas well drilling through shallow layers of groundwater near the surface, traveling about twice as deep as the water table, then turning 90 degrees to create a horizontal lateral. From that horizontal portion extend enormous fractures underground that connect the well to drinking-water sources above. Some renderings include red or black splotches that represent oil, gas, or toxic chemicals moving up into the aquifer. (I was unable to gain permission to use any such images for this book, but a simple online search for "fracking contamination" will turn up plenty of examples.) Any reader presented with these images would understandably worry about fracking fluids destroying the water supply of thousands or even millions of people. In New York City, which draws some of its drinking water from a river basin underlain by the Marcellus—and where drinking water is unfiltered—such an outcome would be ruinous.

But the scales of these images are typically so distorted that they prevent any real understanding of how deep most shale wells go and how far away they are from sources of drinking water. Figure 3.1 gives the actual scale of a typical shale well, this one from the Fayetteville shale found in northern Arkansas. Freshwater resources are found at depths of up to roughly 400 feet (nationally, most private water wells are no more than 500 feet deep, and municipal water wells that help supply drinking water for cities are 1,000 feet deep at most). Companies like Southwestern Energy, which has drilled thousands of wells into the Fayetteville shale, are required by the state to add multiple layers of steel and cement to a depth well below potential drinking-water sources (all states have regulations like this, as described in chapter 6). The purpose of this "surface casing" is to make sure that the stuff inside the well doesn't enter the groundwater.

About 4,000 feet below the drinking water sits the Fayetteville shale, where fracking will take place. How deep is that? Consider New York City's iconic Chrysler Building, which is about 1,000 feet tall. In essence, fracking takes place roughly four Chrysler Buildings below where any drinking water sources exist. Within these 4,000 feet are hundreds of millions of tons of rock, many types of which are impermeable (in other words, fluids and gases won't move through them).

Other major shale plays lie similarly far below any drinking-water sources, including the Bakken (8,000 to 12,000 feet), Marcellus (5,000 to 9,000 feet), and Eagle Ford (4,000 to 13,000 feet). According to one study, the average depth of hydraulic fracturing taking place between 2010 and 2013 was 8,300 feet.[1] In most cases, the length of the cracks in the shale created by fracking is shorter than 200 feet. Very large fractures can

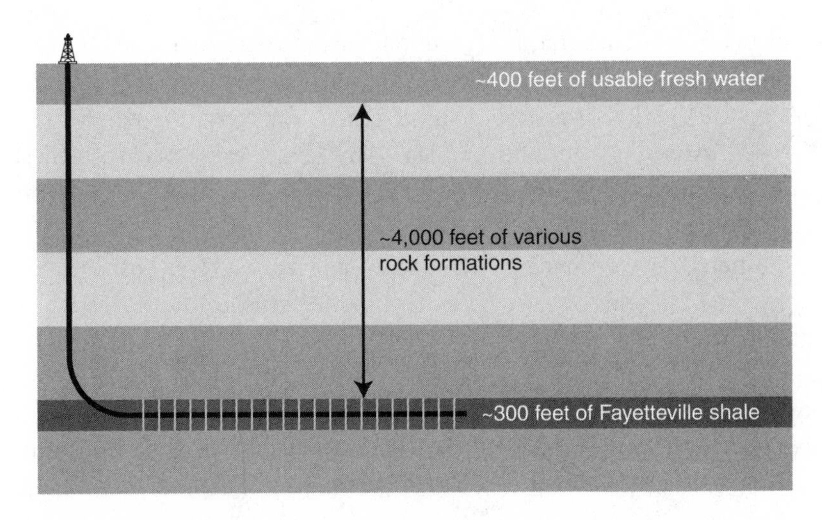

FIGURE 3.1 **A scale-appropriate figure of drilling and fracking**

Source: Adapted from Southwestern Energy. Used with permission.

extend as far as 500 feet or more,[2] still far shorter than would be necessary to connect the shale layer to a water source near the surface.

Is it impossible that fracking fluids could come up to the surface or water table from those depths? No, but it's very unlikely. The brine found in numerous rock layers deep below the surface can naturally migrate from these depths over millions of years[3]—but even this is very uncommon. Indeed, there's a lot of really ugly stuff that deep in the earth's surface, in some cases including radioactive compounds, but those things don't rise high enough to contaminate our water in any systematic way. There's no reason to think fracking chemicals would be any different.

Dozens of research papers have examined drinking-water quality in regions with extensive oil and gas development, and with a single exception (which I discuss later in this chapter), none have shown that fracking chemicals have migrated from deep underground into drinking-water sources.

But if fracking didn't cause those people's water to catch on fire, what did?

FLAMING FAUCETS

For a city dweller like me, water supply is rarely a major concern. My city pumps it from somewhere, treats it, sends it to me, and I drink it, bathe in it, and wash with it. But in rural regions, the process isn't so simple. Many residents need to drill and maintain their own wells or work with neighbors to develop and maintain a steady and safe water supply.

Most oil and gas development takes place in rural areas, and in some cases, private water wells can end up fairly close to new oil and gas wells (states often regulate the minimum distance between these two types of wells, ranging from fifty to one thousand feet). In at least several hundred cases,[4] this proximity has created substantial problems for families reliant on well water (the precise number of incidents is unknown because of opacity from state agencies and sealed settlements between companies and landowners). As I described earlier, the risk of contamination is not primarily from the fracking process itself, since shale formations are typically separated from drinking-water sources by thousands of feet of rock. But there are other risks.

Companies are required by states to add rings of steel and cement to an oil or gas well as it passes through a drinking-water zone. The idea is that this casing creates multiple layers of protection between the stuff traveling through the well—fracking fluids going down and oil, gas, and water coming up—from the freshwater near the surface. If the steel and cement are properly constructed, tested, and maintained, these protections are very effective. However, with tens of thousands of new oil and gas wells drilled each year (in 2010, the most recent available year, roughly 37,000 new oil and gas wells were drilled in the United States),[5] the odds are high that something, somewhere, will go wrong.

The leading problem—which has the *potential* to lead to the flaming faucets seen in Weld County, Colorado—comes from what's often called "stray gas." Stray gas is natural gas, mostly methane, which uses an oil and gas well as a conduit to migrate toward the surface. In most cases, stray gas comes from somewhere in the region between the bottom of the well and the groundwater—it rarely comes from the shale formation or other target formation into which the well has been drilled. But understanding how stray gas can move from underground and into groundwater is complex and requires a bit of an explanation.

THE IMPORTANCE OF INTEGRITY

Thinking back to figure 3.1, there are typically thousands of feet and many layers of rock located between the surface and a shale formation. Although the shale formation is the layer that holds the biggest prize for an energy company, there may also be smaller pockets of oil and/or gas found in the interim layers.

When a company—say our fictional Rollins Energy—moves in to develop that shale layer, it acquires the land, builds a well pad, and starts bringing in equipment to drill and hydraulically fracture the well. Rollins follows the relevant state regulations, adding multiple layers of protective steel and cement around the well to protect water sources. Figure 3.2 illustrates the three layers of steel casing with the labels "conductor pipe," "surface casing," and "production casing."

In this illustration, the drilling and fracking don't cause any problems, and everything goes according to plan: natural gas (represented as little bubbles) from the shale layer, labeled as the "target producing zone," is happily flowing to the surface. (This figure, in order to show details of what's happening in a particular underground location, is not to scale.)

Figure 3.2 also shows that there is another naturally occurring pocket of natural gas (again, little bubbles) in one of the layers that Rollins drilled through, labeled as the "shallow producing zone." But that natural gas stays in place because the steel and cement are doing their job: making sure that the gas, oil, and water outside of the well don't make their way inside.

But wells aren't always constructed properly, and if there are problems with the cement and steel casing, Rollins's well could end up damaging the groundwater near the surface. Figure 3.3 illustrates one of the ways this can happen. In this example, the cement installed by Rollins (or the contractor they hired) does not seal properly. In oil and gas terminology, the space between the steel pipe and the earth is called the annulus. The cement is supposed to seal off the annulus from the rocks surrounding it,

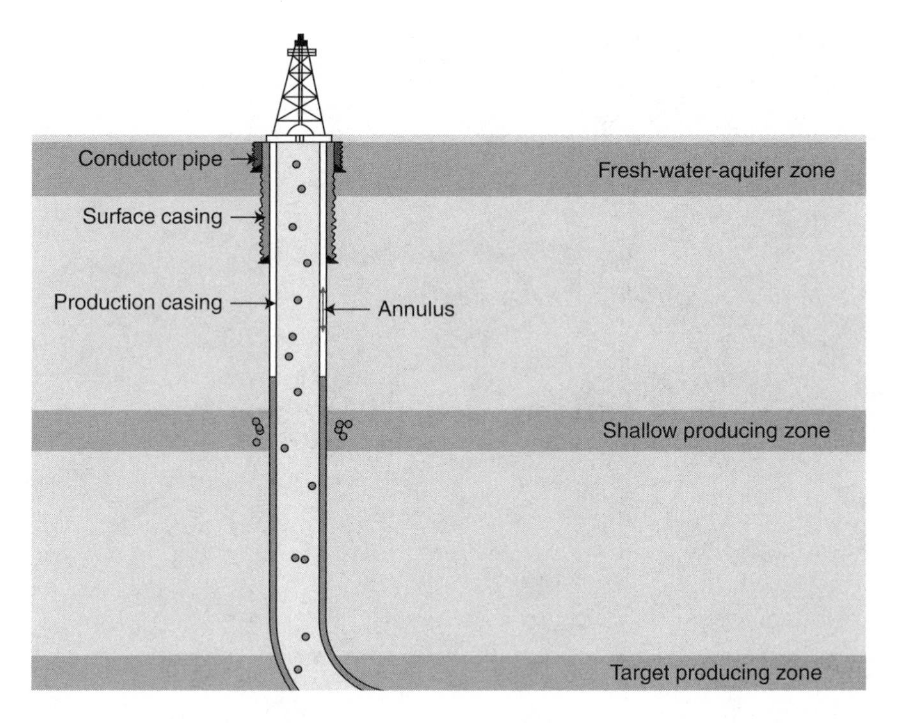

FIGURE 3.2 **A well with good mechanical integrity**

Source: Adapted from Southwestern Energy. Used with permission.

preventing any fluids or gases from using the well as a conduit for traveling up toward the surface. But that's not always what happens. As the figure shows, defects in the cementing have allowed natural gas from the shallow producing zone to enter the annulus.

Because it's now stuck inside the annulus, this gas from the shallow producing zone doesn't go directly into the water table. Instead, it travels up the annulus toward the surface, where pressure begins to build between the rings of steel pipe that were originally designed to protect the fresh-water zone. After enough gas builds up (this could occur over several hours, days, weeks, or months), it starts to seep out from underneath the layers of steel, entering the water supply, as figure 3.4 shows.

This phenomenon of natural gas entering groundwater sources through defects in a well is known in the industry as stray gas, or "methane

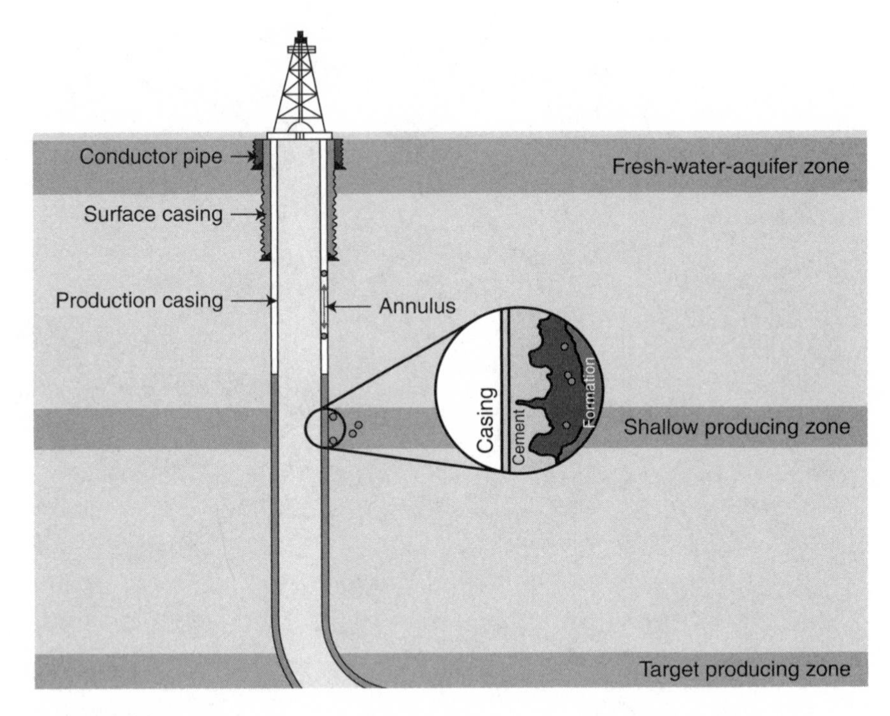

FIGURE 3.3 **Methane enters the well annulus via "cement channeling"**

Source: Adapted from Southwestern Energy. Used with permission.

migration." The route I describe is not the only one through which natural gas can accidentally enter groundwater sources. There can be other types of problems with the cement or defects in the steel casing that could allow natural gas to infiltrate groundwater.[6]

But regardless of how it ends up there, if enough methane gets into the groundwater that supplies a water well, that methane will come out of the tap. And while methane itself is not hazardous to human health, it is odorless, tasteless, invisible, and highly flammable, which means that—under the right set of circumstances—it could accumulate and explode. Hence the flaming faucets.

In 2011, researchers from Duke University published the first in a series of widely cited papers that found high concentrations of methane in drinking water from a Pennsylvania region with lots of Marcellus shale

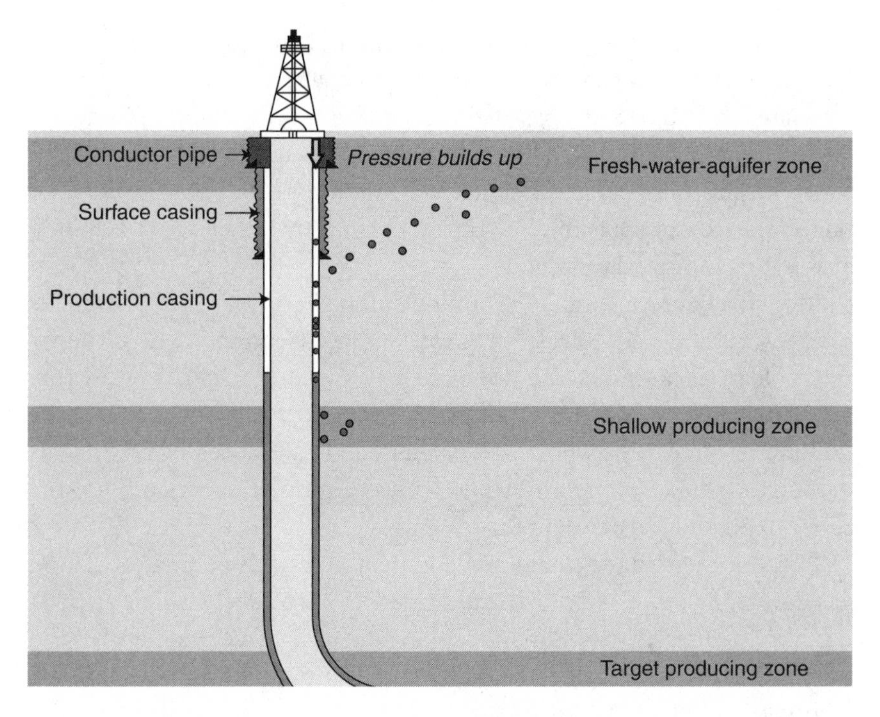

FIGURE 3.4 Methane builds up and escapes into groundwater

Source: Adapted from Southwestern Energy. Used with permission.

development. Although their methods were not definitive (they did not measure methane levels in these wells *before* natural gas development began), they were very suggestive of stray gas.[7] Other papers have followed from these, and other researchers have documented cases of methane migration caused by problems with well casings.[8]

These Duke papers, along with subsequent findings by other researchers, have been the primary academic sources that antifracking movements have seized on, declaring that "fracking poisons water!" But that conclusion is oversimplified. First, methane in drinking water is not new. Numerous reports from the U.S. Geological Survey describe methane in water wells occurring naturally in places such as New York, West Virginia, and more. A 2000 USGS publication states: "Reports from the 1800s document gas bubbles in water wells, in streams, and in fields after heavy rains; this evidence suggests that [methane] migration has always existed."[9] Similar stories from New York State and Pennsylvania exist, including an "eternal flame" burning in a stream about fifteen miles south of Buffalo.[10] Press reports also describe homeowners in the region lighting their tap water on fire in previous decades as something of a party trick.[11] But none of these cases has to do with fracking or any other activity associated with gas or oil development.

Methane in groundwater isn't unique to the East Coast, either. Studies released in 2014 and 2016 from Weld County, Colorado, where the famous flaming faucet in the documentary *Gasland* occurs, found that methane in groundwater was common throughout the region, and while there were some cases where gassy water had been caused by oil and gas drilling, methane in most water wells was naturally occurring, not the result of oil and gas development.[12]

Put simply, methane can wind up in drinking water because of oil and gas development *and* from natural causes. Unfortunately, when discussing flaming faucets, partisans on either side of the fracking debate tend to pick one of these explanations and stick with it, rather than acknowledging that both are possible.

IT CAN HAPPEN: FRACKING CHEMICALS IN THE WATER

Studies conducted in many oil- and gas-producing regions have shown that stray gas can be a real risk of oil and gas development. What they have *not* found is the most common fear related to fracking: that fracking chemicals will migrate from shale formations and up into drinking water. But the world is not a simple place. There are a small number of cases around the country where researchers and regulators have found evidence that fracking chemicals may have infiltrated drinking-water sources. These cases have been quite isolated, affecting a very small number of people. But they should not be ignored. And for industry and regulators, these cases reinforce the fact that fracking has the potential to affect drinking-water sources.

A 2015 paper documents one of several cases in Bradford County, Pennsylvania, where stray gas entered the drinking-water supply (Bradford County sits just next to Susquehanna County, where Dimock Township is located). But the paper found more than just methane in the water: it showed that a small, nonhazardous amount of a compound called 2-n-butoxyethanol, a chemical used in fracking fluids, had also made its way into the groundwater. How this compound ended up in the water is not certain, but the authors hypothesize that there may have been a natural fracture between an uncased portion of the well and a family's water supply.[13] Again, a well-construction problem, but this time, one that resulted in contamination by fracking chemicals. 2-n-butoxyethanol is only toxic at much higher levels, but this paper reports a serious and important finding: risks to groundwater may in some cases include fracking chemicals, reinforcing the importance of good well construction.

Another seminal case involves the town of Pavillion, Wyoming, home to about two hundred people along Wyoming's beautiful Wind River mountain range. Drilling and fracking had been going on in the area for some time, with over one hundred natural gas wells drilled just to the east of Pavillion from 2000 to 2008. In 2008, several local residents contacted the U.S. EPA, complaining that their water had acquired a foul odor and bad taste, which they attributed to the nearby gas wells. The EPA launched an investigation. Unlike most fracking operations, hydraulic fracturing was taking place in Pavillion at surprisingly shallow depths. One well had been fractured at just 1,220 feet, within striking distance of local drinking-water supplies.

In December 2011, the EPA released a preliminary report suggesting that hydraulic fracturing itself, along with leaks from nearby wastewater pits, had contaminated local wells with a variety of chemicals.[14] In early 2016, a peer-reviewed study of the case agreed that fracking fluids and wastewater were indeed the likely culprits.[15] However, the Wyoming Department of Environmental Quality, which took over the official investigation from the EPA in 2014, a few months later issued their final report, which stated that oil and gas activities (including fracking) were *not* the likely cause of water contamination in the area.[16] Uncertainty remains.

A third case documenting potential effects on groundwater from fracking comes from a study carried out in West Texas' Permian basin region.[17] In the study, researchers looked at well water supplying homes in an area where dozens of new oil and gas wells were being drilled and fracked. One of the key advantages of this study was that it looked at the water before, during, and after many of the wells were drilled, measuring any changes in quality along the way.

The researchers tested the wells at four intervals over the course of roughly one year. During the first test, the quality of the water was fairly good—a normal pH level and no pollutants associated with fracking or other contamination from oil and gas development. But during the next two tests, as drilling and fracking increased in the area, several troubling compounds appeared in the groundwater, including methanol and dichloromethane, both of which can be harmful to human health. Methanol is also a common component in fracking fluids. Other indicators of water quality, including the pH level and the amount of total organic carbon in the water, spiked.

But at the fourth test, just several months later, many of these pollutants had disappeared. The researchers couldn't explain the changes and couldn't tell whether oil and gas operations were the cause. They concluded that we may know even less than we think about changes in groundwater quality: why different chemicals appear at different times, why some seem to come and go at will, and how resilient groundwater sources are to these temporary contamination events.

THE INDUSTRY KNOWS!

In his 2011 short film *The Sky Is Pink*,[18] Josh Fox shows several slides and papers from oil and gas presentations and journals that depict how errors in well casing and cementing can lead to stray gas. The film uses the documents to argue that oil and gas wells are "inherently contaminating," causing damage to groundwater at alarmingly high rates. Fox interprets one industry report to claim that "over a thirty-year period, 50 percent of well casings failed" and cites others that purportedly show "enormous" and "astronomical" rates of wells leaking methane and other potential contaminants.

Like James Inhofe's argument that fracking has never polluted drinking water, Fox's claim too has some basis in reality. However, Fox's interpretation of the studies is, at best, overly simplistic, and the conclusion that viewers are asked to draw is highly misleading.

Stray gas and the issue of properly cementing oil and gas wells are indeed a challenge in the oil and gas industry. But they are hardly a secret. Claiming otherwise, Fox compares this research to studies conducted by the tobacco industry, arguing that oil and gas companies were keeping damning evidence "in their drawers" to prevent public understanding of risks. He cites as evidence slides from presentations by executives at Southwestern Energy showing a variety of risks from oil and gas development. However, these slides have been shown publicly by Southwestern to academics and community groups for years and were provided by the company for use in this book (see figures 3.2 through 3.4). The explicit purpose of the presentation is to educate audiences about the risks of water contamination. Hardly the stuff of conspiracies.

More substantively, Fox cites a study from an industry journal, making the claim that 50 percent of well casings fail. The authors of that study examine wells in the Gulf of Mexico and make a subtler point: that a large percentage of these wells exhibit something called sustained casing pressure (SCP), an indicator that some amount of gas or liquids has made its way from outside the well and into the annulus. However, this is a far cry from indicating well "failure." Indeed, the authors note later in the paper that in some oil- and gas-producing regions, SCP is present in every well, but in less than 1 percent of those wells do gas or fluids move to the surface, where environmental damage might occur.[19]

Hundreds of papers on this topic are available in petroleum-engineering journals. States have extensive regulations detailing what type of cement to use, how deep it must be placed, how long it must sit, what pressure it must withstand, and so on. The problem has been documented in the real world plenty of times. For example, part of the cause of BP's disastrous 2010 Macondo well blowout, explosion, and oil spill was improper cementing.[20] In the BP case, the consequences were enormous and long lasting. In other cases, cementing failures that result in stray gas can be fixed, eliminating the problem.

The real question is how often these cases of stray gas happen and how often they affect households and drinking-water supplies. While we don't have great answers to these questions, a few studies shed useful light on the issue. At one end of the argument, a 2014 paper led by the professor and antifracking advocate Anthony Ingraffea argues that cement failure in new wells is widespread and more common than in older, nonshale wells.[21] The paper finds that, across Pennsylvania, cementing and casing problems appear to be present in roughly 2 percent of wells (other statewide studies find between 1 and 3 percent).[22] More troublingly, they find that in some regions of the state, closer to 9 percent of wells show defects. But these findings come with a fairly important caveat: they identify instances of cementing problems, not cementing problems that resulted in stray gas or other pollution.

How often do these problems result in actual contamination? The Pennsylvania Department of Environmental Protection's annual oil and gas report provides such numbers. As figure 3.5 shows, the number of stray-gas cases has generally declined since 2010, despite a large number of new wells drilled across the state. For every new well drilled, the number of stray-gas cases has declined from 0.8 percent in 2010 to 0.23 percent in 2014 and 0 percent in 2015.

Unfortunately, most states don't regularly publish these kinds of data, so there are no equivalent nationwide estimates. But there are studies that provide additional context. In one international study, two petroleum engineers note that the failure of any single layer of cement does not mean that environmental damage will occur.[23] They describe how failures are not uncommon in certain parts of the world, especially where wells are older, poorly maintained, or subject to natural hazards like major earthquakes. For example, one study of oil and gas wells in and around the Straits of Malacca (a narrow Pacific Ocean waterway between Indonesia and Malaysia and an area of substantial recent seismic activity) found that 43 percent of wells had some degree of casing problems. However, just 1 to 4 percent of wells had a plausible pathway to leak and cause environmental damage.[24]

Closer to home, a 2011 study examined tens of thousands of regulatory records in two states (Ohio and Texas) to estimate how groundwater contamination can occur around oil- and gas-production sites. Just 0.06 percent and 0.02 percent of wells in Ohio and Texas, respectively, experienced failures in casing or cementing that resulted in groundwater contamination.[25] One more study from Colorado estimated that the likeli-

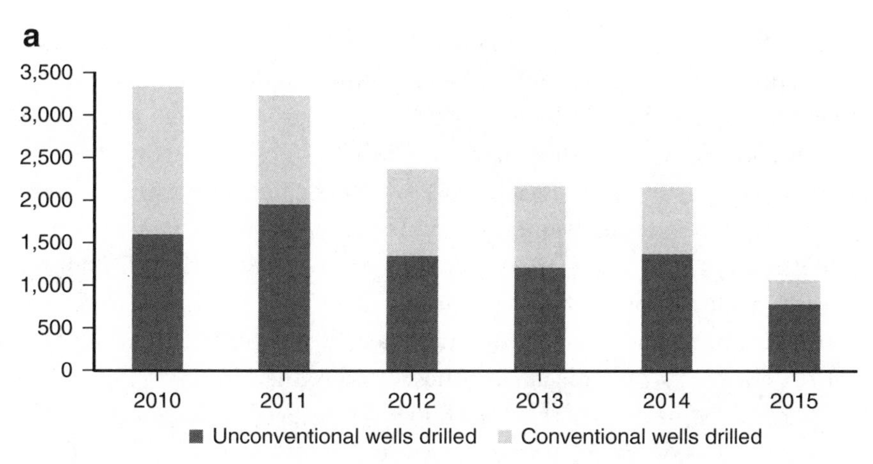

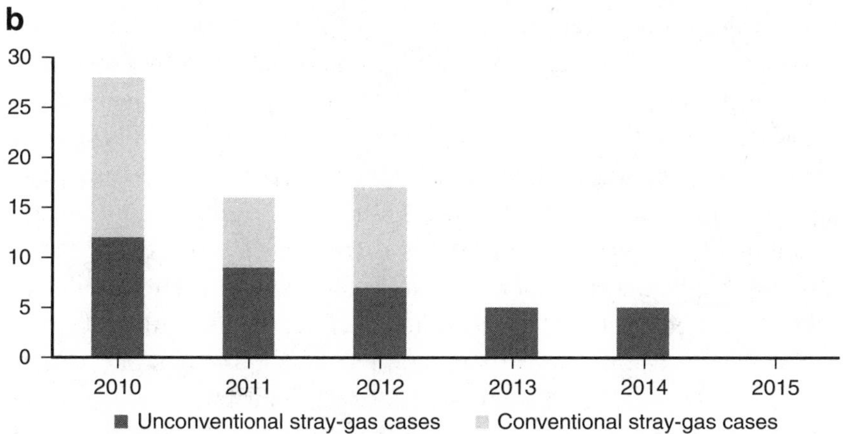

FIGURE 3.5 **Wells drilled and stray-gas cases in Pennsylvania**

Source: Pennsylvania Department of Environmental Quality, "2015 Oil and Gas Annual Report," http://www.dep.pa.gov/business/energy/oilandgasprograms/oilandgasmgmt/pages/annual-report.aspx.

hood of oil and gas wells failing, potentially allowing methane to migrate into groundwater, was somewhere between 0.06 and 0.15 percent.[26]

So where does that leave us? First, we know that the issue of stray gas is not new, we know it can be caused by poor well construction, and we know that it can create real risks to groundwater and the residents that rely upon it.

It's also important to note that the transportation, refining, and storage of oil and its refined products like gasoline and diesel are in fact more likely to pollute groundwater than oil and gas production (e.g., drilling and fracking). A 2015 report from a consortium of Texas state government agencies, universities, and NGOs shows that the number of new groundwater-contamination cases in Texas fell from over 1,000 in the late 1990s to under 300 in 2014 and 2015, even while oil and gas drilling surged. Indeed, most cases of groundwater contamination were not caused by oil and gas production but instead by leaking gasoline- or diesel-storage tanks, oil and gas pipelines, or oil refineries.[27]

Here's one final way of thinking about the risks of fracking to groundwater: if oil and gas drilling and hydraulic fracturing could result in large-scale groundwater contamination—the type of contamination that would affect an entire city—wouldn't we know about it by now? More than one hundred years after the great Spindletop gusher, about 120,000 people live in Beaumont, Texas. The city pulls some of its drinking water from the Chicot aquifer (part of a larger system called the Gulf Coast aquifer) and some from the Neches River. A 2015 sampling of the city's water quality revealed no violations of the EPA's water-quality standards,[28] despite the fact that the geological formation where Spindletop's oil was trapped lies inside the boundaries of the Chicot aquifer.[29]

In Oil City, Pennsylvania (population 10,000), more than 150 years after the United States' first commercial oil well came into production and where thousands of wells were "stimulated" by the Roberts Petroleum Torpedo Company, the city pulls its drinking water from a series of local groundwater wells. According to 2014 testing,[30] water quality met all relevant standards with one exception: the presence of coliform bacteria, a contaminant that can come from human or animal waste and that can be found in plants or soils—but does not come from oil and gas activities.

SPILLS AND WASTEWATER

For most experts who study the industry, work for energy companies, or regulate those companies, a more pressing concern relates to the potential for spills of hazardous chemicals and the handling of wastewater from oil and gas wells.[31]

Spills are easy to understand. Imagine a truck hauling several large containers of potentially hazardous chemicals to a well site for fracking. If that truck crashes, there is a potential for contamination and environmental damage. If workers at the well site mishandle some of these chemicals or store them improperly, there is another risk. For example, a 2015 study sampled water wells in northeastern Pennsylvania and found that surface spills of fracturing fluids had seeped down to drinking-water wells, affecting more than half of the sixty-four wells the researchers examined.[32] Importantly, the levels of contaminants found in these water wells were below the thresholds that would constitute a risk to human health. Nonetheless, the findings were concerning.

As with stray gas caused by improper well construction, spills of oil, chemicals, and other hazardous materials are almost inevitable, at least at some scale. There is always the possibility of human error, shoddy enforcement, or companies that cut corners regardless of the regulations. For example, in 2015 a pipeline near the coast of Santa Barbara ruptured, spilling more than 100,000 barrels of crude oil into the surrounding environment. The state of California charged the company with a number of infractions and sought to impose fines beyond what the company had already paid in cleanup costs.[33]

These types of large spills do not generally come directly from oil- and gas-production sites but from leaks at larger storage sites or spills from pipelines. In broad terms, the number of oil spills in the United States has increased as the shale revolution has led to the installation of new pipelines and increased volumes flowing through older lines. However, the *volume* of oil spilled has declined over the same period of time, as shown in figure 3.6.

A more complex issue relates to handling oil and gas wastewater. Water used in fracking that returns to the surface is called "flowback," and the naturally occurring water that flows up with oil and gas is called "produced water." Produced water can contain brine, radioactive elements, and all sorts of other toxic stuff that occurs naturally thousands of feet underground. When this water comes to the surface, handling it, treating it, and disposing of it can cause some substantial problems.

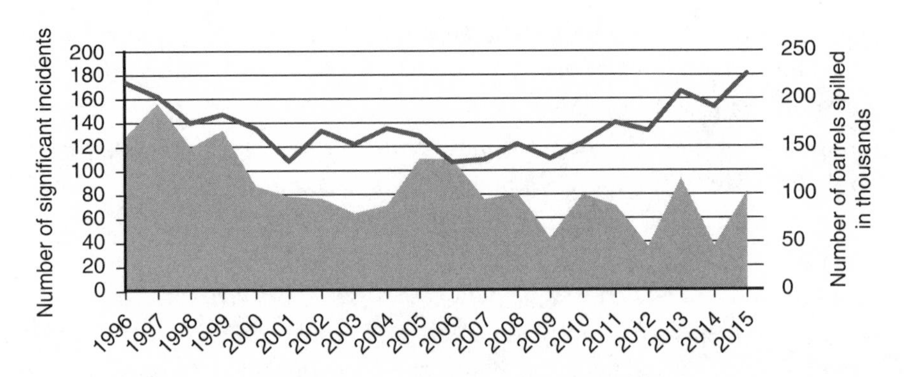

FIGURE 3.6 U.S. hazardous-liquids significant incidents and spill volume

"Significant incident" is defined as any incident that includes a fatality or injury requiring in-patient hospitalization, $50,000 or more in total costs (measured in 1984 dollars), highly volatile liquid releases of five barrels or more or other liquid releases of fifty barrels or more, or liquid releases resulting in an unintentional fire or explosion. Shaded areas of the figure indicate barrels spilled; the line shows significant incidents.

Source: U.S. Department of Transportation, Pipeline and Hazardous Materials Administration, "Data and Statistics," https://www.phmsa.dot.gov/pipeline/library/data-stats.

For example, before wastewater is collected and hauled away, it's often collected in storage ponds, which are usually dug next to well sites. In most states, these ponds are required to be lined with impermeable plastic, which should prevent contaminated water from seeping into the soil below. But if those storage ponds are improperly constructed or if the plastic lining tears, wastewater may spill over the top of the pond or sink into the surface below.

There are numerous confirmed cases of leaking wastewater-disposal pits causing problems,[34] and the issue may have been responsible for the high-profile case of groundwater contamination in Pavillion, Wyoming, which I described earlier in this chapter.[35] In two Pennsylvania cases, companies have been fined millions of dollars because their wastewater-storage ponds polluted local water sources.[36]

An analysis of twenty-six years of data from Ohio and sixteen years from Texas sheds some light on the hazards of wastewater management compared with drilling and fracking.[37] Based on data from state regulators, the report identifies 185 incidents of groundwater contamination in

Ohio and 211 cases in Texas. In both states, the most common culprit for these incidents was issues related to the handling of wastewater, either at the well site or during the wastewater-disposal process. The state records showed no cases where fracking was directly responsible for groundwater contamination.

Most states employ a host of regulations in the effort to prevent such spills and leaks, but a couple have relatively weak regulations, and enforcement of those rules can be spotty (I explore regulatory issues in depth in chapter 6).[38] Some states have moved toward requiring companies to store their wastewater in tanks, which reduces the risk of water spilling over the top but increases other risks, like a tank failure caused by a flood, tornado, or lightning strike.

After leaving the well site, wastewater is usually transported by trucks or pipelines. In regions where trucks are used, the massive volume of water produced from the well (sometimes ten or twenty barrels of water are produced for every barrel of oil) means lots and lots of trucks. This raises the obvious risk that a truck may crash, leaking hazardous materials into streams or other sensitive areas where they can harm local habitats and wildlife. There are numerous examples of these types of spills,[39] though it's difficult to quantify how widespread they are nationally or how severely they have affected habitats.

Trucks hauling wastewater usually travel to one of three locations: first, they may go to a recycling facility, which treats the water and returns it to the oilfield for use in another frack job. This is fairly common, and some companies invest millions of dollars in recycling facilities. I visited one such site in Garfield County, Colorado, where a company built miles of pipelines to run their wastewater from well sites to a centralized water-recycling plant.

A second option is that trucks or pipelines transport the wastewater to a disposal well. These wells, which have gathered attention largely because of their connection with earthquakes (which I discuss in chapter 5), are a long-time method of disposing unwanted oil and gas wastewater and are regulated by the U.S. EPA under the Safe Drinking Water Act of 1974.[40] For many years, disposal wells have generally been a safe and effective method of getting rid of waste products from oil and gas wells and other industrial facilities. However, there have been cases of real concern, and injecting wastewater into the wrong place can cause substantial risks.

Wastewater wells generally pump water down into rock formations far below the water table, sometimes into old oil and gas reservoirs. Similar to the injection of fracking fluids into the earth, these wells are drilled deeply enough that there is little risk of this waste seeping up toward the surface and affecting drinking water.

But if wastewater-disposal wells are pumping into *useable* or *drinkable* sources, you've got a problem. Such a problem was uncovered in 2015 in central California, where outdated regulations and monitoring protocols appear to have allowed some disposal wells to inject wastewater into aquifers that could be future sources of water for irrigation or human consumption.[41] In the context of a historic drought, such an oversight is hard to understand and represents a substantial failure of policy implementation and regulatory oversight. A few months after reporting outlets uncovered the story, press reports indicated that California regulators had shut down the problematic disposal wells.[42]

In several instances, wastewater-disposal sites have been linked with contamination of nearby streams, and press reports have described cases of wastewater wells failing.[43] If this issue were widespread, it would be troubling indeed, as there are roughly 180,000 of these disposal wells around the United States.[44] Thankfully, they don't seem to be common, the pollution is not widespread, and there don't appear to be any immediate risks to public health. Still, the fact that it is happening at all should make regulators and the industry look closer at just how safe these disposal methods really are and ensure that regulations are designed and implemented properly.

For wastewater transported through pipelines, there are also risks. In 2016, a news report and a separate academic study looked at wastewater spills in North Dakota (some from pipelines, others from trucks) and found that spills resulted in the increase of a variety of contaminants, some of which persisted in soils for years after the initial accident.[45] This type of contamination can in most cases be cleaned up by removing the soil, but the costs and time associated with that cleanup can be substantial, not to the mention the disruption and economic losses it can cause to landowners and nearby residents. And as with just about every other environmental risk, rules and regulations are effective only if they are enforced.

For example, a farmer named Daryl Peterson, who describes himself as supportive of the oil and gas industry ("I'm not a tree hugger"), has been featured in numerous press reports stating that wastewater spills have taken a substantial toll on his North Dakota farm. Most of the oil wells on the family farm are decades old, and wastewater spills from old pipelines on the property had never been cleaned up properly. As a result, large patches of the land are barren, and other patches grow only weeds. As of late 2016, Mr. Peterson's farm had yet to be properly restored.[46]

Finally, wastewater has in the past been trucked to nearby municipally owned wastewater-processing facilities, the types of plants that treat sewage and other household wastes. In most cases, asking these facilities to treat oil and gas wastewater is a bad idea. During the early years of Pennsylvania's shale gas development, a substantial share of wastewater was transported to these facilities, treated, and then released into nearby streams and rivers. While this process of treating and releasing wastewater is routine for households, businesses, and other common sources of waste, it is not designed to handle the extremely salty, chemical-rich water flowing from oil and gas wells. As a result of this improper disposal method, water samples in parts of Pennsylvania indicated high levels of pollutants, including salts and radium (a naturally occurring radioactive element present in some produced water), downstream from these wastewater facilities.[47] This method of water disposal and resulting contamination generated substantial alarm.[48]

But after realizing that municipal treatment plants were not the right places to manage oil and gas wastewater, the state of Pennsylvania worked with oil and gas operators to stop this method of disposal. And in 2016, the EPA prohibited companies from disposing of their wastewater at these types of treatment plants unless it was cleaned up ahead of time.[49] As a result, this problem appears to be largely dormant and has not arisen as a major issue in other parts of the country. It has, however, offered a lesson on how *not* to manage wastewater. And importantly, no evidence has suggested that the temporary mistreatment of Marcellus wastewater has caused long-lasting contamination to Pennsylvania waterways.

SMALL SCALE VERSUS LARGE SCALE

Having read through this chapter, you are likely thinking about all of the risks I've just described. You may have thought about additional risks related to water that I didn't describe in detail. And while I've emphasized that the risk of each particular incident is very small, you may be wondering something along the lines of: "With all these risks combined, isn't it inevitable that fracking will cause widespread environmental damage and adversely impact public health?"

The key word in this question is "widespread."[50] As you consider the issues described in this book, one of—if not *the* most important—ideas to keep in mind is *scale*. With the hundreds of thousands of existing oil and gas wells in the United States and the additional tens of thousands that are drilled and fracked each year, it is virtually inevitable that some scale of environmental damage will occur.

But how large are those risks, and how widespread are the damages when they occur? Based on existing evidence, it looks as though fracking is unlikely to cause water contamination at a scale that would affect entire cities. After over one hundred years of companies drilling and blowing stuff up underground, the evidence of contamination affecting thousands or millions of people simply isn't there. Public water supplies in places like Beaumont or Oil City have not seen the type of citywide pollution and health risks experienced in, for example, Flint, Michigan, where, for reasons having nothing to do with oil and gas extraction, lead-tainted water has caused widespread suffering.[51]

But water contamination has happened at smaller scales and is likely to continue as long as drilling takes place close to the places where people live. Oil and gas workers make mistakes. Oil and gas companies sometimes cut corners. Regulators, too, make mistakes or can be overstretched and understaffed. Regulators may also be too accommodating of the industry they are supposed to oversee, a major finding in the aftermath of the Deepwater Horizon disaster and subsequent oil spill in the Gulf of Mexico (notably, there was no hydraulic fracturing done at that site). Each of these facts illustrates that drilling wells, fracking them, and moving oil and gas around the world—like most other aspects of life—is an exercise in risk management.

Back in Dimock, fracking is still a controversial issue, but not for the reasons you might guess. While several homeowners along Carter Road whose wells were contaminated with methane continue to oppose fracking vocally, a larger number of residents are upset about a *lack* of drilling. The Pennsylvania Department of Environmental Protection, after fining the company that caused the stray gas, imposed a moratorium on new drilling within the township. As a result, dozens of homeowners who had expected large paydays from royalties have waited for years for their fracking ship to come in. The elected township officials that I met with in Dimock described this lack of drilling as their largest concern over fracking. The end result? An uneasy truce in a pastoral corner of northeastern Pennsylvania, with no resolution in sight.

SUMMING UP

If you live far from oil and gas activity, fracking will not affect your water supply. If you live near an oil and gas field and get your water from a private well, there is a small chance your water could be negatively affected by oil and gas development. The primary risk is not from fracking chemicals, which generally stay far below groundwater levels, but from natural gas, which can escape either directly from the well through faulty steel and cement casings, or from naturally occurring gas closer to the surface that travels to water sources through well defects. There are also risks to local water sources from surface spills of oil, chemicals, or salty brine. This pollution has occurred hundreds of times in recent years, but in the context of the tens of thousands of wells drilled each year in the United States, the risks are relatively small.

4

WILL FRACKING MAKE ME SICK?

Driving through Karnes County, Texas, it's impossible to miss the signs of the shale revolution. The land is flat, and with few trees to impede the view, well sites—indicated by orange natural-gas flares, oscillating steel pump jacks, and large beige storage tanks—pop up at regular intervals. Karnes County sits atop the Eagle Ford shale, which spans thousands of square miles and forms a crescent moon south of San Antonio. From 2005 through 2016, the number of wells producing from the Eagle Ford formation has grown from less than 10 to more than 16,000.[1]

Unlike the rolling hills and emerald pastures of northeastern Pennsylvania, the Eagle Ford feels like oil country. And if Dimock, Pennsylvania, represents the risks of oil and gas development to local groundwater sources, parts of the Eagle Ford have come to embody the broader health risks of oil and gas development.

In 2014, the Weather Channel, Inside Climate News, and the Center for Public Integrity released a joint reporting project focused squarely on this issue.[2] Reporters visited homes in Karnes County and attended public meetings as residents complained about the impacts to their health from air emissions at well sites. One resident, Lynn Buehring, told reporters that more than fifty wells had been drilled within 2.5 miles of her home and explained that while she had experienced allergies before, new symptoms that she blamed on Eagle Ford development were far worse: "There's this heaviness on my chest and it feels like an elephant sitting here, and like somebody's choking the air out of me, and I can't get a breath." More than most oil and gas plays, Eagle Ford wells produce

FIGURE 4.1 A well site in Karnes County, Texas

Source: Author, January 2017.

hydrogen sulfide, a poisonous gas with a sweet odor that can cause headaches, nausea, and—if experienced in an acute dose—"nearly instant death," according to the U.S. Occupational Safety and Health Administration.[3] At some Eagle Ford locations, stark warning signs caution curious visitors as they approach the well site: *Danger: Poison Gas.*

Not all oil and gas wells produce "sour gas," as the mixture of natural gas and hydrogen sulfide is known. But all produce at least some potentially harmful air emissions. Some of these emissions occur during drilling, others during fracking, and others after the well has been completed. As shale development has increased, so have questions over the health risks it poses, and researchers have moved in to try to determine the extent of these risks.

Antifracking groups have claimed for years that the health risks are large. A simple Internet search for "fracking" and "cancer" turns up a variety of news stories, op-eds, and images from protests heralding a link between the two. Headlines from such a search include: "Toxic Chemicals,

Carcinogens Skyrocket near Fracking Sites" and "Fracking Chemicals Linked to Cancer, According to New Study."

So how accurate are the headlines, and how well founded are the fears? What are the air emissions that might make people sick? What about fracking fluids? Are they toxic, and, if so, how likely are they to make us sick? What other risks might oil and gas development pose to nearby residents and communities? And what questions remain unanswered?

CONCERN OVER TOXIC CHEMICALS

One of the most persistent fears of fracking relates to the numerous chemicals that are blended with water and sand to fracture shale formations deep underground, allowing oil and gas to flow into the well and up to the surface. While industry advocates point out that water and sand usually account for 98 to 99 percent of the mixture used in a typical hydraulic-fracturing treatment, the total volume of fluids can range from 1 to 20 million gallons (though usually at the lower end of the spectrum).[4] That means the remaining 1 to 2 percent of fluid could represent anywhere from 10,000 to 400,000 gallons of chemical additives. These numbers are hardly trivial: a typical gasoline tanker truck carries anywhere from 5,000 to 10,000 gallons of fluid.

The list of chemicals that may be used in any single frack job stretches into the hundreds. From this menu of options, a smaller number of chemicals is chosen for any given fracturing treatment, with selections made based on the specific characteristics of the targeted rock formation and a variety of other factors.

Why are chemicals used at all in the process? If water alone were used to fracture the rock, several problems could emerge that would reduce production from the well and cause other issues. Here are a few basic examples of the types of compounds that may be added to improve the performance of the well:

After the rock is fractured and oil and gas begin to flow out from the formation, fragments of rock may come loose and impede the progress of the oil and gas moving toward the surface. To counteract this effect, some sort of acid (such as hydrochloric acid) can be added to dissolve those compounds, making it easier for the oil and gas to flow out of the rock and into the well. Since these acids can corrode the steel pipes that make up the well, another compound is often added to prevent the acid from damaging the well itself. Importantly, the acids used for fracking do not travel far from the bottom of the well—rather, they encounter rock, dissolve it, and are then neutralized.

Another common addition is a biocide (such as glutaraldehyde), which kills bacteria that may be present in the fresh water used for fracking, or bacteria that flow into the well from the rock formation along with the produced water, oil, and gas. If they're not killed off, these bacteria could potentially corrode the well or inhibit the flow of oil and gas toward the surface.

A friction reducer is often added to make the water "slick." As I described in chapter 2, the first successful frack jobs on shale formations originally developed in the Barnett were called "slickwater" hydraulic fracturing. While water might seem to be pretty slick already, the intense heat and pressure of shale wells means that even water needs a little extra lubrication. (Imagine how a spilled glass of water makes your kitchen floor slippery, then imagine how slippery a spilled bottle of motor oil would be, and you can get the idea.)

These are just a few examples of the purpose chemicals serve in fracking. The simple point is that each chemical plays a role in making the well more productive. Detail on the types of chemicals added to fracking fluid are available through sources such as fracfocus.org, a website where companies disclose the chemicals they use for each well that is hydraulically fractured.

The major concern is whether these chemicals will cause harm to humans. Researchers have looked at this question from a variety of perspectives. First, a number of studies have tested the health effects of exposure to one or more of the chemicals used in the fracking process and found, unsurprisingly, that direct exposure to a sufficient amount of these chemicals can harm animals and people.[5] In one notable study, researchers found that mice exposed to a variety of chemicals used in fracking experienced endocrine disruption,[6] which can lead to negative developmental and neurological effects, particularly for newborn or *in utero* babies. Other studies have similarly found that exposing animals to relatively high concentrations of fracking chemicals causes negative reproductive or other health outcomes (hence colorful headlines such as "Fracking Could Give You Massive Balls, Tiny Sperm Count").[7]

While it's true that direct exposure to these chemicals can be harmful, the results of these experiments don't actually tell us how risky fracking is. In large enough quantities, all sorts of fluids and chemicals that we interact with every day—gasoline, medicines, household cleaning products—can be harmful to humans and animals.

Instead of asking whether fracking chemicals cause harm in large doses, we instead should ask: are humans being exposed to fracking chemicals in quantities sufficient to cause harm? A close look suggests the answer is probably no. The main reason that health damage is unlikely is that, while some fracking chemicals can be harmful if encountered in large quantities or over long periods of time, the likelihood of people who live in fracking areas coming into this type of contact is extremely small. To date, there is no research that indicates that the health of people living near oil and gas wells has been—or is likely to be—harmed by exposure to the chemicals mixed in with fracking fluid. As I described in the last chapter, there are only a few documented cases where fracking fluids may have migrated into drinking-water sources.

Instead, human or environmental exposure to these chemicals is more likely to be caused by something else: a truck accident, well blowout, or spill on the well pad. Indeed, the greatest risk of exposure to these chemicals is among the oil and gas workers who handle the chemicals, not the general public. And while risks for industry workers are relatively well understood (as I'll discuss in a moment), there is much less information that helps us understand risks to the general public.

For example, no studies have systematically gathered data on fluids spilled from trucks working with hazardous fracking chemicals, but there are data on the broader transportation industry. Nationally, there are roughly 5,200 truck accidents involving hazardous materials each year, out of a total of more than 800,000 such shipments. On average, 36 percent of those accidents lead to spills.[8] That works out to roughly a 0.2 percent chance that a hazmat truck's trip will result in a spill.

While this is a low probability, the large numbers of wells and trucks spread across the country means that some spills are likely to occur somewhere. If roughly twenty truck trips are required to carry chemicals to each frack job[9] and roughly 25,000 oil and gas wells are hydraulically fractured in the United States each year, we would expect fifty-nine annual spills from trucks attributable to fracking (though the sizes of these spills are still uncertain).

For context, consider the enormous fleets of eighteen-wheelers that every day haul millions of gallons of gasoline, diesel, and other hazardous materials around our cities, towns, and rural landscapes. In 2016, the U.S. Pipeline and Hazardous Materials Safety Administration reported almost 16,000 accidental releases of hazardous materials on U.S. highways.[10]

The effects of each spill can be severe for a limited area, but society generally accepts these risks because of the benefits that gasoline and other potentially harmful substances provide us. We do our best to prevent accidents from occurring, but when they do occur, we respond as quickly as possible. The same risks, which are real but also well understood and manageable, apply to fracking chemicals.

Well blowouts are also uncommon, but they can have an even greater effect. They occur when the pressure inside an oil or gas well becomes too great for the steel and cement to withstand, and the driller loses control of the well (the 2011 Deepwater Horizon blowout in the Gulf of Mexico is perhaps the best-known example). Blowouts can spew a variety of chemicals, including oil, brine, and other fluids, onto the surface. They pose a health risk not just to the workers on the well site but also to nearby residents.

For example, a well blowout in eastern Ohio's Utica shale region lasted for ten days, forcing the evacuation of twenty-eight local families.[11] Another blowout in the same region led to a large fire that engulfed several onsite chemical-storage tanks; contaminated a stream, killing an estimated 70,000 fish; and forced the evacuation of nearby residents.[12] National statistics on blowouts are hard to find, but in Texas, where 98,000 wells were drilled between 2011 and 2015,[13] there were 104 blowouts or other well-control problems,[14] translating to 0.1 percent of wells. Importantly, blowouts can happen at any well, whether it is fracked or not.

Clearly there are some real risks from high-impact accidents that, though very rare, could expose people and the environment to fracking chemicals. But are there other health risks from oil and gas development more broadly? The answer is yes: while the pumping of chemicals deep into shale formations looks highly unlikely to result in widespread health risks, other processes at or near oil and gas well sites raise important concerns.

WHAT ELSE COULD MAKE YOU SICK?

Oil and gas development—especially the large-scale, multiwell operations that increasingly characterize modern shale plays—involves lots of heavy machinery. Earth movers such as bulldozers, graders, and large trucks hauling rock work together to prepare the well site, emitting diesel fumes along the way. Once the well pad is prepared, a drilling rig comes on site to drill wells—sometimes a single well, in other cases ten wells or more on a single pad. The drilling rig also typically runs on high-powered diesel generators, as do the various pieces of equipment that monitor and assist the well's progress.[15]

Once the wells are drilled, more eighteen-wheel diesel-powered trucks haul water, sand, and chemicals to the well site, a process that can involve hundreds or even thousands of round trips (in some locations, temporary pipelines transport water to and from the site, reducing truck traffic substantially). At the same time, diesel generators power the high-pressure pumping trucks that perform the fracking itself—pushing the water-based mixture into the wellbore at 10,000 pounds per square inch or more.

In short, shale development is an industrial operation—albeit a small one compared to a steel mill or oil refinery—that involves heavy machinery mostly powered by diesel engines. For workers in the oil and gas industry, it's not news that oil and gas production is an industrial process. Indeed, oil and gas production is one of the most dangerous

FIGURE 4.2 High-powered diesel engines powering a hydraulic-fracturing operation

Source: Joshua Doubek, via Wikimedia Commons, "Fracking Operation," August 11, 2011, https://commons.wikimedia.org/wiki/File:Fracking_operation.JPG.

occupations in the United States. On a per-worker basis, it was the third most fatal occupation in 2015, behind agriculture/forestry/fishing/hunting and transportation/warehousing,[16] and workplace injuries are a constant concern and major focus of the industry.[17]

But one of the hallmarks of shale development is that it has occurred in places where oil and gas drilling had either never happened before or had only happened at a small scale. In Fort Worth, Texas, southwestern Pennsylvania, and in the suburbs north of Denver, shale development sometimes occurs a few hundred feet from homes, businesses, and, occasionally, schools.[18] This proximity to population centers, while not entirely new to the industry (for example, oil drilling has occurred for decades in Los Angeles), raises questions about how people might be exposed to air emissions from drilling, fracking, and other phases of oil and gas development.

One major issue of concern is the use of all those diesel engines. These engines emit volatile organic compounds (VOCs), particulate matter, carbon monoxide, and other potentially harmful air pollutants. As wells are developed over a period of months, these emissions persist in quantities that vary depending on the intensity of industrial activity at the well site. Will people exposed to these emissions throughout the construction, drilling, and fracking processes get sick? It's possible. As the U.S. EPA describes in a variety of reports, emissions from diesel engines are carcinogenic to humans if exposure is high enough over a long period of time (months or years). Diesel emissions also cause irritation and inflammation in short-term acute doses.[19]

Along with being emitted from diesel engines, VOCs are often emitted during "flowback," the stage after the rock has been fractured and the fracking fluids, oil and gas, and a variety of other compounds are flowing back to the surface. Flowback can also produce air toxics such as benzene or hydrogen sulfide, both of which are known to be harmful to humans[20] and which come to the surface along with produced oil and gas. While about 75 percent of VOCs are emitted by natural processes (mostly from trees) and also come from household products such as glue, paint, and permanent markers, the oil and gas industry is the largest anthropogenic emitter of VOCs in the United States. The EPA estimates that in 2014, about 19 percent of total anthropogenic emissions, or roughly 3.2 million tons of VOCs, came from oil- and gas-production activities.[21]

VOCs are a leading factor in the formation of smog, which exacerbates asthma, leads to increased hospital-admission rates, and can contribute to premature death for sensitive populations.[22] While very small doses of these VOCs are extremely unlikely to be harmful (think of the fumes you inhale when filling up your car's tank with gasoline), those who are exposed to them on a recurring basis, such as oil- and gas-industry workers, are at greater risk. According to the National Institutes of Health, short-term exposure to VOCs can result in "eye and respiratory tract irritation, headaches, dizziness, visual disorders, fatigue, loss of coordination, allergic skin reactions, nausea, and memory impairment." Longer-term exposure can result in "damage to the liver, kidneys, and central nervous system," and two VOCs, benzene and formaldehyde, are "human carcinogens."[23]

During flowback, the fluids, such as oil and water, are captured, and the gases are often burned in "flares," the bright orange flames that flicker next to many oil wells. Flares convert the raw natural gas to CO_2 and water. However, the flare may fail to convert 100 percent of the gases flowing back, and some gases may be vented rather than burned. During flowback, which often lasts about a week (though this can vary depending on several factors), pollutants that could harm human health are flowing into the air.

A number of studies have documented air quality at or near oil- and gas-development sites, showing elevated levels of these potentially harmful emissions. Similarly, the presence of diesel exhaust from onsite generators and truck engines is well documented.[24] Regulations in states such as Colorado and Pennsylvania have required companies to capture most of these emissions by using "green completions," which means capturing most or all of the produced natural gas, rather than venting or flaring it, and virtually eliminating VOC emissions during flowback. This positive step has also been mirrored voluntarily by some companies.[25]

Along with regulations, another approach to identifying and reducing risks has come from the state government of Colorado, which established in 2016 a program for residents to report their oil- and gas-related health concerns. These reports are tracked by the state's Department of Health and Environment.[26] If the state receives complaints that suggest potentially harmful emissions from a well pad, they send out a crew with specialized equipment to test local air quality. And in 2016, the U.S. EPA proposed regulations requiring all new oil and gas wells to use green completions, which—if the regulations are implemented by the Trump administration, a highly uncertain prospect—would reduce these air emissions and with them the risks to populations living near new oil and gas wells.

So it is clear that there are potential risks to human health from emissions associated with oil and gas development. Antifracking advocates often note the emissions I describe and assume that the negative health effects associated with them will inevitably occur. But we need to ask: How are humans exposed to these emissions? What is the frequency (how often), duration (how long), and intensity (how much) of exposure? Frustratingly, high-quality research on these questions is limited (and low-quality research is much easier to find), but I'll describe some key studies in the following section.

WHAT DOES THE RESEARCH SAY?

Roughly ten years into the shale revolution, the state of research on health effects from fracking and from oil and gas development more broadly is quite limited. A 2017 review of the research, coupled with more than 10,000 data points collected by the Colorado Department of Public Health and Environment, found that evidence to date does not warrant immediate additional action to protect public health. It does, however, state that because of the paucity of information, more research of higher quality is needed.[27]

This uncertainty has not prevented breathless headlines in press outlets (particularly the left-leaning ones) trumpeting studies purporting to show substantial negative health effects from shale development. And while a number of researchers have examined the question, a variety of issues makes it difficult to come to a broad conclusion.

One ever-present challenge for researchers in any field is distinguishing between correlation and causation. While correlation shows that two outcomes occurred at the same time (a new oil well was fracked in my backyard, and at the same time, I got a nosebleed), causation establishes that one event actually led to the other (a new oil well was fracked in my backyard, which caused my nosebleed). To distinguish between correlation and causation, researchers have devised a set of procedures to isolate the root cause of a particular outcome.

If we were magically to remove all practical and ethical constraints from doing research with human beings, studies of health effects would employ randomized controlled trials, which randomly select individuals to be part of a "treatment" group or a "control" group. Researchers would take detailed measurements of participants' health, establishing a baseline to measure any future changes. The treatment group would be exposed to oil- and gas-development activities, perhaps by going to live near an active well site. The control group would be given some sort of placebo; for example, perhaps they would be sent to live near a fake oil and gas site, where they believe they are exposed to oil and gas production but are not in fact faced with any of the risks. This perfect experiment would involve a large number of participants, allowing researchers to identify the major effects of receiving the treatment rather than the control.

Since the logistics and ethical pitfalls of such an experiment are prohibitive, researchers take different approaches to approximate this gold standard of randomized controlled trials. While no approach is perfect, some studies have done better than others at disentangling correlation from causation. However, the relatively small amount of research that we have on this topic makes it difficult to offer tidy or definitive conclusions.

Some early studies on the health risks of oil and gas development reported negative effects from living near production sites, and while these studies have received substantial attention, they rest on methods that are, in some cases, highly suspect. Specifically, data are gathered through interviews or surveys with people who volunteered to report health impacts of shale development.[28] Because the studies rely on self-reported outcomes, it is unclear whether the participants experienced real changes in health conditions or whether they simply *believe* that they have experienced changes. In addition, the researchers have no way of identifying whether oil and gas development or some other cause is the driver of any perceived changes in health. What's more, responses to these types of surveys and interviews will be biased because they are not random. That is, the results are skewed because people who believe they have been affected are more likely to participate than those who perceive no effect.

Other studies have been carried out more carefully but are still limited by a variety of factors. For example, one study carried out around Rifle, in western Colorado, examined water samples to identify whether they contained elevated levels of compounds that could affect estrogen and androgen receptors in humans, which could in turn lead to adverse reproductive-health outcomes.[29] The study, which was carried out by a group of researchers at the University of Missouri, reported elevated levels of these compounds near natural gas production sites, which in turn led to headlines such as "MU Researchers Find Fracking Chemicals Disrupt Hormone Function."[30]

However, there are a number of reasons to be cautious when interpreting the study's findings. First, the researchers did not perform baseline testing, meaning they did not know about local water quality before drilling occurred. Second, they collected only forty-two samples of water at six sites near well pads and just five samples from two other locations, to serve as a basis for comparison. Any conclusions would be far more reliable if they rested on results obtained from hundreds or thousands of samples from dozens or hundreds of sites. These limitations aren't necessarily the fault of the researchers. Like everyone else in the field, they work with limited time frames and limited budgets, making it impossible to conduct the ideal experiment. Nonetheless, these methodological shortcomings mean that headlines proclaiming what appear to be straightforward risks often mask a great deal of uncertainty.

Another study from the same region examining the health of newborns offers mixed results, demonstrating how difficult it can be to draw clear conclusions about health risks.[31] The study uses a large dataset and finds a statistically significant correlation between babies born to mothers living close to well sites and increased instances of congenital heart defects. This troubling finding led to headlines such as "New Study Links Fracking to Birth Defects."[32] However, that same study also found a statistically significant (but smaller than the other finding) correlation between babies born to mothers living near well sites and *decreased* preterm birth (a positive health outcome). Unsurprisingly, there were no headlines trumpeting: "New Study Links Fracking to Healthier Babies." The discrepancy in headlines says more about the media environment than it does about the quality of the researcher's work. Indeed, the authors note in this study that it's hard to draw simple conclusions from their results and highlight the need for more data collection and analysis.

As I discussed earlier in the chapter, it is well known that air emissions from oil and gas production can be harmful if encountered in sufficient quantities. Recent papers have reestablished this finding by measuring emissions at or near well sites in a variety of locations.[33] Other papers have gone a step further, trying to tease out the potential health effects of those emissions. One 2012 paper looks at exposure to benzene and other potentially harmful compounds in the same region of western Colorado as the Rifle study, estimating that those living within half a mile of a well site were at a higher risk of health impacts, including cancer, than those living farther away.[34] They estimate that the cumulative cancer risk for those living within half a mile of a well was ten in a million, compared with six in a million for those living farther than half a mile from a well. However, another 2012 study conducted by the Colorado Department of Public Health and Environment found that benzene emissions from natural gas drilling sites were well within the limits established by the U.S. EPA.[35]

In another confusing pair of studies from the same researchers, a 2017 paper found that Colorado children living close to a large number of oil and gas wells were more likely to be afflicted by a certain type of cancer than those who lived farther away (they examined two types of cancers and found effects for only one). However, the study was limited by several factors, including a small sample size, not knowing what type of activity was occurring at the well site, and not knowing whether the mothers of these children were smokers.[36] A study released around the same time by the same Colorado agency examined 10,000 air samples from oil- and gas-producing regions and found no measurements where harmful substances including benzene or formaldehyde were above "safe" levels, even for vulnerable populations such as the elderly. It also reviewed the existing literature and found "no substantial or moderate evidence for any health effects," though—much like the findings described in this chapter—there's still a lot we don't know.[37]

Another study examined five states, finding that air quality near some oil and gas production, processing, and distribution facilities included benzene, formaldehyde, and hydrogen sulfide at levels that could cause acute health effects.[38] The results showed that 20 out of 76 samples included potentially harmful levels of certain compounds and included some (again, self-reported) health effects for individuals at those locations, such as nosebleeds, coughing, and watering eyes. However, the study did not collect baseline data, which again makes it hard to know whether oil and gas operations were the root cause of these emissions and resulting health impacts.

A smaller number of studies have used more rigorous methods to test the health effects of shale development. This work has two key advantages relative to most of the studies previously discussed. First, they directly measure health outcomes, rather than measuring air quality or water quality, which may or may not actually affect people. In addition, those health outcomes are not self-reported but instead are based on data from hospitals or other health providers. Second, these studies use large anonymous datasets, which makes it more likely that the results will be representative of the broader population. While these studies also have limitations, they offer perhaps the most compelling evidence that oil and gas development may have some real risks for those living or working nearby.

One study from southwestern Pennsylvania indicates that children who were carried *in utero* within areas of dense drilling activity (more than six shale wells per square mile) had a lower birth weight and higher incidences of SGA (small for gestational age) than those living in areas with less drilling activity.[39] The authors, however, note that while the difference in birth weight is statistically significant,[40] it is small, meaning that the effects of this outcome are not entirely clear. Another study examining health effects for children carried *in utero* found that preterm births were more likely when mothers lived closer to shale activity. Importantly, this study did not find negative health effects on a variety of other infant-health metrics, such as low birth weight, SGA, and other metrics.[41]

A 2015 study in northeastern Pennsylvania found that people living in counties with high amounts of fracking activity ended up going to the hospital for cardiac issues more than people living in one county without fracking,[42] and a 2016 study in the same region found a correlation between living close to a new shale well and the onset of asthma attacks, which can result in hospitalization.[43] But like much work in complex areas of public health, these studies do not claim to have found a causal link between shale development and negative health outcomes. In other words, the negative health outcomes they found could be caused by something other than shale development, primarily because definitively identifying the precise cause of any health outcome is extremely difficult.

While the existing body of research does not provide "smoking-gun" evidence of major health effects from oil and gas development, it certainly raises serious concerns. Because these studies do not definitively establish causation, proindustry advocates sometimes dismiss them, implying that we shouldn't worry about the health impacts of fracking and of oil and gas development more broadly.[44] But that's the wrong lesson to draw. Instead of dismissing concerns, policy makers need to take into account the evidence that does exist when they consider the costs and benefits of regulating the industry. And perhaps more importantly, additional research is needed to provide a better understanding of the scale and scope of the risks.

THE HEALTH BENEFITS OF DISPLACING COAL

One clear health *benefit* of the shale revolution is the reduced level of coal-fired power production brought about by lower natural gas prices. Compared to natural-gas-fired power plants, coal plants emit far more harmful—in some cases, carcinogenic—pollutants.[45] These emissions include mercury, particulate matter, and other air toxics harmful to human health.

While coal has played an enormous role in advancing human well-being by powering the industrial revolution in the United Kingdom, United States, and other nations, its negative health effects have been enormous. For example, London and the industrial cities of the United Kingdom have struggled to manage smoke from coal fires ever since the 1300s. As coal supplied more and more energy to the city—heating homes, powering factories, and generating electricity—London's air quality continued to deteriorate, eventually reaching a nadir over several days in 1952: the Black Fog (also known as the Great Smog or the Big Smoke).

Because of an unusual temperature inversion, London's streets were shrouded in fog so dense that vehicles slowed to a crawl, and even an opera had to be cancelled when fog crept into the theater. Over the next several days, smoke and dust from coal fires turned the fog from gray to black, limiting visibility to just eleven inches. Hospitals began to fill with people complaining of respiratory ailments. By the time the fog cleared four days later, an estimated four thousand Londoners had been killed by the smog, many found laid out on city streets or in parks.[46]

While the Black Fog illustrates the extreme end of coal's downsides, its negative health effects are rarely so dramatic or easily attributed. Like research on the health effects of oil and gas development, calculating the health effects of coal-fired power plants is difficult and subject to uncertainties. But the estimates that do exist do not bode well for the black stuff. One study released in 2000 (when coal provided a larger share of the United States' electricity) estimated that emissions from all domestic power plants were responsible for roughly 30,000 premature deaths per year, and the large bulk of these deaths were associated with emissions from coal-fired plants.[47] Because of ensuing federal regulations that reduced some of the most harmful emissions, a more recent estimate—which came from the aptly named report "The Toll from Coal"—put the number of premature deaths at 13,000, attributing them entirely to emissions of fine particles from coal-fired power plants.[48] Indeed, the toll from coal grows further when one adds negative health outcomes that do not result in premature death, such as additional hospitalizations, asthma, and lung disease.

Economists have attempted to quantify the damages associated with burning coal and other fossil fuels. In a wide-ranging 2011 study that calculated the damages from a variety of activities, including mining, fuel processing, power production, and more, the authors estimate that for every dollar of value coal-fired electricity adds to the economy, it causes roughly $2.20 in air-pollution damages, including the negative health effects of air pollution. They estimate that every dollar of value that natural gas–fired power adds to the economy is only accompanied by $0.34 in damages.[49]

Experts will debate whether these estimates are too high or too low, but if they are anywhere close to accurate, it's clear that public health has benefited from the displacement of coal by natural gas. This positive outcome is often overlooked because the benefits of cleaner air are shared widely across large swaths of the country, whereas oil and gas extraction affect a concentrated group of people living near well sites. It's much easier to tell a story about a family that has suffered negative health effects from shale development than it is to tell a story about the tens or hundreds of thousands of people who have now avoided the harmful emissions from coal plants.

Of course, renewable sources of electricity such as wind and solar have also displaced coal in recent years, and they deserve some of the credit for improving air quality and public health. However, as I discuss in more detail in chapter 7, the largest cause of decreased coal-fired power has been cheap natural gas, and while some advocates argue that these renewable sources can today replace both coal *and* gas,[50] most experts believe that renewables are not yet up to the task and that natural gas can play a useful role over the next several decades by further displacing coal-fired electricity.[51]

SUMMING UP

Existing research on the potential health effects of shale development is limited, and the research we do have does not paint a simple picture. To date, we lack definitive answers about what the biggest risks are, how long they may last, and what types of illnesses may be caused. Like the issue of potential water contamination, additional and higher-quality research is needed to make a better assessment of the extent of these health risks.

While there do not appear to be massive or widespread risks to large numbers of people (for example, entire towns or regions being negatively affected), there is reason to be concerned about those who live or work near oil- and gas-production sites. Anecdotes—backed by limited research—suggest that air emissions from well sites may lead to some risk of issues such as eye, ear, nose, and throat irritation and possibly more serious long-term conditions if exposure takes place over a period of months or years. Research also suggests reason for concern over babies carried *in utero* near new oil and gas wells.

Looking forward, regulations of these emissions, along with robust enforcement, would go a long way in reducing the health risks posed by oil and gas development (more on regulations in chapter 6). At the same time, natural gas' displacement of coal has already had widespread health benefits that need to be taken into account.

For some groups, any health risks from shale development will be too great. But like many other complex environmental issues, oil and gas development presents a difficult exercise in weighing the pros and cons, the risks and benefits—and there are few easy answers.

5

DOES FRACKING CAUSE EARTHQUAKES?

In the one-stoplight town of Anthony, in central-southern Kansas, people were starting to get worried. Cracks were showing up in the walls, and the daily shake had long since stopped being an amusing curiosity.

In March 2015, I traveled to meet with local government officials not just in Anthony but in the small towns of Medicine Lodge and Pratt, just a few miles north of the Oklahoma border. I had been on the Oklahoma side a few months earlier for meetings in Alva, Cherokee, Enid, and Medford. The land is spectacularly flat, and spotting towns from a distance is easy. Each city—especially those on the Kansas side—features an enormous white grain elevator located somewhere near its center, towering above the rest of the landscape. The elevators resemble collections of overstuffed drinking straws, all bunched together and gleaming white, often rising more than three hundred feet high, distributing wheat from the heartland out and into the world.

Beneath the surrounding wheat and soybean fields that fill the landscape is a rock formation called the Mississippian Lime, between 3,000 and 6,000 feet below the surface. The Mississippian is not a shale but a limestone, and for decades it has produced modest quantities of oil from thousands of wells dotting the border between Kansas and Oklahoma. (It's called the Mississippian in reference to the geological age in which it was formed, not for the state of Mississippi.) But production from the Mississippian has surged since 2010, thanks to the application of horizontal drilling and hydraulic fracturing, the same types of techniques that helped increase production from shale.

Another rock formation in the region, called the Hunton, has also seen growth in drilling over the past decade. And the Hunton, where companies sink both vertical and horizontal wells—but typically do not use hydraulic fracturing—is a particularly prolific producer of salty water. In fact, it is often referred to as the "Hunton Dewatering play" because of the large initial volumes of water that are produced alongside the oil and gas (as noted earlier, this produced water is commingled deep underground with oil and gas). Over time, water production from the Hunton decreases while oil and gas continue to flow.

As in other regions that have experienced growth in oil and gas production, a large number of new wells has meant a large increase in the amount of wastewater, which is generated both from the prehistoric water embedded in the rock, which is liberated along with produced oil and gas, and to a lesser extent the flowback of fracking fluids. In northern Oklahoma and southern Kansas, wastewater from the Hunton and Mississippian plays has grown rapidly, and as discussed in chapter 3, preventing wastewater from contaminating soils or aquifers has been one of the key challenges for both the industry and for regulators.

Beyond risks of contamination to water and land, wastewater has helped create another problem for the oil and gas industry: earthquakes, a topic that has captured widespread media attention and generated a sense of wonder among the general public. How could it be that Oklahoma, which had previously experienced just a couple quakes per year greater than 3.0 in magnitude, had about nine hundred quakes of this size in 2015 (3.0 is the typical threshold for quakes that people can feel at the surface)? By comparison, California had around two hundred in the same year.

The problem of human-caused earthquakes, also known as "induced seismicity," has been concentrated below the wheat fields and grain elevators of northern Oklahoma and southern Kansas, though it has also occurred in parts of Ohio and northern Texas. In the United States, the process of fracking has not been the primary culprit for these earthquakes. There have been a few documented cases in which the hydraulic-fracturing process itself was responsible for earthquakes in Canada, Ohio, Oklahoma, Texas, and England,[1] but these cases are much more rare, have generally been smaller, and have not caused damage to property.

Instead, wastewater produced by the thousands of new wells across the United States has been the leading cause. As larger volumes of wastewater have been injected into disposal wells (special-purpose wells that don't produce oil or gas but are designed to store industrial waste), a small portion of this water has resulted in a startling increase in earthquakes, particularly in Oklahoma.

I'll describe two specific cases later in this chapter, but in general, if wastewater (or any other fluid) is injected into an underground formation so as to alter the pressures along an existing fault, it becomes more likely that that fault will slip, causing a quake. With greater volumes and a greater pressure of fluid comes a greater probability of induced seismicity.

Since around 2010, these quakes have caused property damage, rattled dishes and knocked ornaments from shelves, unnerved homeowners, and imperiled historic structures in mostly rural communities. Regulators in different states have varied in how aggressively they've attempted to stop the shaking, and these varied efforts have resulted in a wide range of successes and failures in managing the problem.

WASTEWATER DISPOSAL

Wastewater from oil and gas wells has been injected into underground rock formations for decades. While the process might at first glance seem risky, longstanding protocols for constructing, maintaining, and operating these wells have generally resulted in the safe, cost-effective storage of waste products from the oil and gas industry as well as waste such as chemicals and other hazardous materials from other industries.

In the 1930s, oil and gas operators in Texas began disposing of the salty produced water that came up along with their oil by pumping it back down into the formation where it originated. Over the following decades, this method of wastewater disposal became more common, spreading to other industries, including refineries and chemical companies. States began developing and implementing regulations to safeguard these disposal wells, but in the 1960s and 1970s, it became apparent that some were causing groundwater contamination, and one in the Denver, Colorado, area was causing hundreds of small earthquakes.[2]

In the Colorado case, wastewater from the Rocky Mountain Arsenal, where the U.S. military had for decades manufactured and dismantled chemical weapons, was injected roughly 12,000 feet underground about six miles north of Denver. Shortly after injecting began in 1962, local monitors began detecting small earthquakes, which became more frequent over the next several years as injections continued. While most quakes were undetectable at the surface, the largest registered at a magnitude of 4.3.[3]

As it turned out, the wastewater was being injected into areas with preexisting faults, which were already lined with fluids. The additional fluid injected by the Rocky Mountain Arsenal reduced friction along those fault lines, allowing the rocks to slip. Although injections were halted in early 1966, the added pressures along these fault lines persisted, and small quakes continued, becoming rarer and rarer until they ceased in the early 1980s.[4]

In 1974, Congress passed the Safe Drinking Water Act, which—among many other things—gave the EPA the authority to regulate these types of wastewater-injection wells. Under the law, the EPA works with state environmental agencies to develop regulations and monitoring programs that are carried out on the ground by employees of the state government. This is called the Underground Injection Control (UIC) program, and while there have been some cases of pollution associated with UIC wells, it has generally done a good job of preventing injection wells from contaminating water or causing other environmental or health hazards.

For the purposes of this discussion, I'll focus on disposal wells, which have been the source of most of the recent human-caused quakes in the United States. Disposal wells are one of four types of wells regulated by the UIC program. They pump fluids far below any drinking-water sources, into rock formations that trap the wastewater and prevent it from moving toward the surface.

As oil and gas development has increased in the past decade, vast new quantities of wastewater have been produced and must be disposed of. Because fracking typically uses millions of gallons of water—and much of that water flows back to the surface during the early days and weeks of production—the drilling of new wells has been one important source of this increased wastewater.

But water from fracking isn't the only important contributor to the growth of wastewater disposal. In fact, it's not even the largest source, as produced water volumes in most regions easily surpass flowback (the fracking fluid that returns to the surface). In some oilfields, fifty to one hundred barrels of water come out of the ground for every barrel of oil. The amount of water produced per barrel of oil is typically called the "water cut," and it ranges from well to well, but every oil and gas reservoir produces at least some water. With the number of new wells drilled in the United States growing by leaps and bounds, the need for wastewater disposal has increased in turn.

For example, in Oklahoma in 2009, producers pulled about 183,000 barrels of oil and about 5.2 billion cubic feet of natural gas out of the ground each day (for an energy equivalent total of about 1.1 million barrels of oil per day). In that same year, they injected about 2.25 million barrels per day of salty wastewater into disposal wells. As oil and gas production started to grow across the state, the volume of wastewater grew, too.[5]

While similar trends can be found in other states, Oklahoma is particularly interesting because the volume of wastewater disposal grew much more rapidly than the volume of oil and gas production. As figure 5.1 shows, the new plays being developed in Oklahoma were particularly water intensive, and most of that new wastewater was being injected into a rock formation called the Arbuckle. Soon it became clear that all of that water going into the Arbuckle was creating some unexpected, unforeseen, and unwelcome consequences.

INDUCED SEISMICITY

It has been known for decades that wastewater disposal and other oil and gas activities could cause earthquakes and changes at the surface. For example, so much oil was produced underneath the city and coastline of Long Beach, California, that some parts of the city sank more than

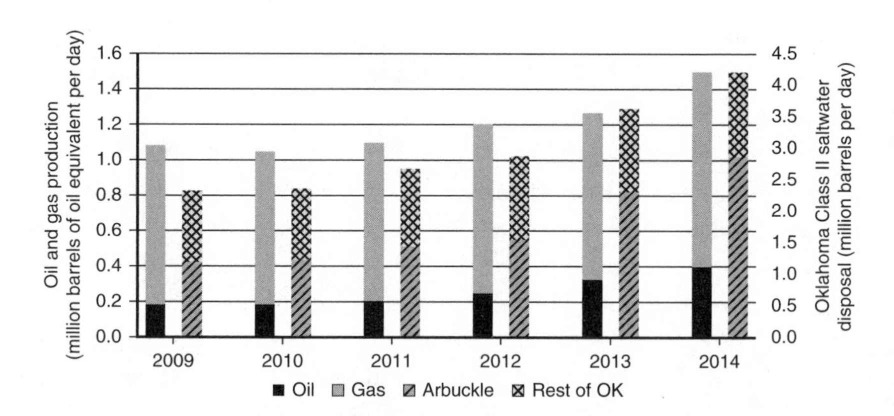

FIGURE 5.1 **Oil production and saltwater disposal in Oklahoma, 2009–2014**

Sources: For oil and gas production: U.S. EIA, "U.S. States," https://www.eia.gov/state/. For saltwater disposal data: K. E. Murray, "Oklahoma Geological Survey Open File Report OF5—2015" (2015), http://ogs.ou.edu/docs/openfile/OF5-2015.pdf.

twenty feet in the 1940s and 1950s. As oil, water, and gas were pumped out from beneath the city, underground pressure was reduced, and the weight of the overlying rock compacted the depressurized area,[6] leading to major property damage (this subsidence was eventually halted by pumping water into the depleted oil reservoirs to maintain pressure). A 2016 paper hypothesized that several large quakes in southern California, including a 1933 Long Beach quake (magnitude 6.4) that killed more than one hundred people, may have also been caused by oil- and gas-production activities, though the connection was not certain.[7] In the southwest, two recent papers examined earthquakes in Texas and Oklahoma going back decades, finding a variety of ways that oil and gas activities may have been the culprits for quakes in those states.[8]

But in recent years, oil- and gas-producing states have unintentionally performed an experiment on the risks of quakes caused by wastewater disposal, with Oklahoma as the best example. Over the past ten years or so, as disposal volumes have surged, the state experienced a rapid growth in the number of earthquakes felt at the surface (figure 5.2).

As I discussed in the previous chapter, correlation shouldn't be conflated with causation. And although most quakes were occurring in re-

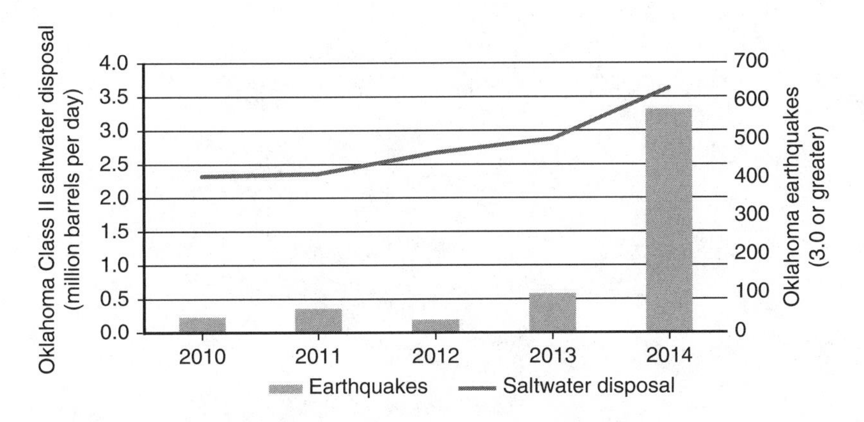

FIGURE 5.2 Oklahoma saltwater disposal and earthquakes

Sources: For saltwater volumes: K. E. Murray, "Oklahoma Geological Survey Open File Report OF5—2015" (2015), http://ogs.ou.edu/docs/openfile/OF5-2015.pdf. For earthquakes: U.S. Geological Survey, "Search Earthquake Catalog" (2016), https://earthquake.usgs.gov/earthquakes/search/.

gions with lots of wastewater disposal, researchers in Oklahoma and at universities around the United States were not initially sure that wastewater was the cause of the rapid growth in quakes.[9] However, over several years of research, it became clear that wastewater injection was indeed the culprit, both in Oklahoma and in other regions that had recently experienced unusual seismicity.[10]

Eventually, researchers concluded that as more and more wastewater was pumped into the Arbuckle, some of that water was beginning to migrate to the "basement rock" below. Similar in some ways to what happened at the Rocky Mountain Arsenal, these fluids found their way to preexisting faults, which were already subject to stress from multiple directions (see figure 5.3). The addition of more water altered the pressures along the fault lines, making it more likely that those rocks would slip past one another, resulting in an earthquake.

The vast majority of these quakes have been small and harmless, causing neither damage to property nor threats to people. Modern-engineered buildings are generally not at risk of structural damage at magnitudes of less than 5.[11] However, older or poorly constructed buildings have experienced substantial damage, power for thousands of people has been

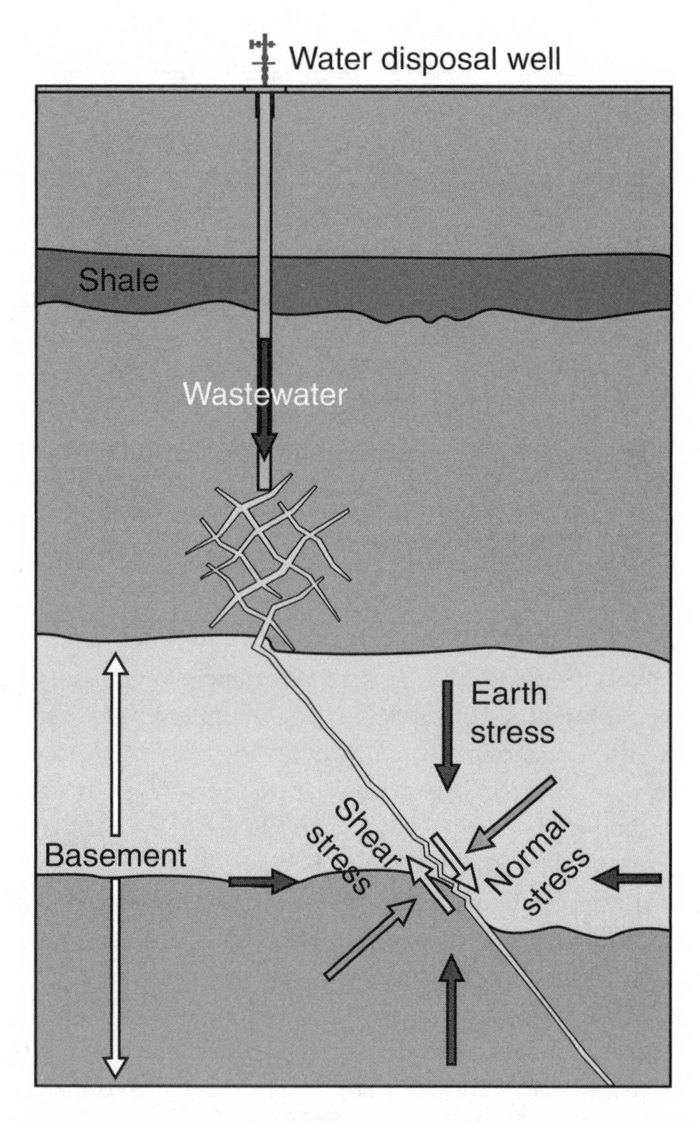

FIGURE 5.3 Wastewater injection and connection with faults in basement rock

Source: Adapted from Southwestern Energy. Used with permission.

knocked out by some recent induced quakes,[12] and nonstructural damage (such as dishes falling from shelves) affects residents and, potentially, causes property values to drop. And while most of Oklahoma's quakes have been small, three have surpassed the 5.0 level: a 5.6 quake in 2011 along with a 5.0 and a 5.8 in 2016. News reports from the area show that these larger quakes have been responsible for damage to structures in and around Oklahoma's cities and rural areas.[13]

The dramatic increase in Oklahoma quakes has led some media outlets to refer to Oklahoma as the new "earthquake capital of the United States," taking that undesirable crown from California.[14] However, because the quakes in Oklahoma have been smaller than those experienced in California, the amount of seismic energy released in California's quakes has been much greater. As a result, the potential for damage has been—and continues to be—much higher in California.

Indeed, it is the intensity of the quakes, more than the sheer number, that determines the amount of damage that may be caused. For example, in 2014 Oklahoma had 580 earthquakes above a 3.0 magnitude on the Richter scale, while California had 265. However, one of those California quakes registered at 6.8 and another at 6.0, while the largest Oklahoma quake was 4.4. The Richter scale is logarithmic: a 6.0 quake releases more than thirty-one times as much energy as a 5.0 quake. Thus the quakes in California were far more powerful and dangerous than anything Oklahoma has experienced in recent years. Put another way, the cumulative amount of energy released from Oklahoma earthquakes from 2009 through the end of 2016, if released in a single event, would have a magnitude of about 6.1.[15]

But as I traveled across northern Oklahoma and southern Kansas, these facts did not offer much comfort. People in California expect earthquakes, but people in Kansas and Oklahoma don't. Almost everyone I asked had a story about a friend or neighbor who suddenly noticed cracks in their walls, had pictures fall from shelves, or noticed an eerie swaying while lying in bed on the second story of their home. At one county courthouse in southern Kansas, a government official warned me that a recent inspection had indicated that the 150-year-old courthouse might be structurally unsound. I laughed nervously and looked for the nearest exit.

WASTEWATER OR FRACKING? WHY THE DIFFERENCE MATTERS

Predictably, pro- and antifracking partisans have done all they can to obscure any nuances surrounding the issue of whether it is wastewater disposal or hydraulic fracturing that causes earthquakes. To be sure, both processes involve pumping large volumes of water deep underground, but the policy response to quakes caused by fracking would be very different from the response to quakes caused by wastewater disposal.

Antifracking groups go straight to headlines such as "Confirmed: Oklahoma Earthquakes Caused by Fracking," when the much more important issue is wastewater disposal.[16] Prodrilling groups trumpet that "Fear of Fracking Earthquakes Is Misplaced," which, strictly speaking, is correct but misleads readers by failing to explain that the cause of the quakes is, in fact, still connected with oil and gas activities.[17]

And why should we care about which terms are used in the fracking debate? As I discussed in chapter 2, mistrust between advocates on either side of the debate is rife, and a good portion of that mistrust is likely attributable to arguments that fail to recognize the validity of the other side's claims. As the extremity of each group's position increases, it becomes harder to find agreement on any solutions that minimize the risks and maximize the benefits of the shale revolution.

For example, if a substantial portion of the public believes that hydraulic fracturing is responsible for the rapid growth of quakes in a place like Oklahoma, they are more likely to push for a blanket ban on fracking. If policy makers act based on this public pressure (as they have in places including New York, France, and Germany) and prohibit hydraulic fracturing, the vast majority of shale plays will not be developed. As a result, the volume of wastewater produced from oil and gas wells will decrease, limiting the future risks of earthquakes. While the goal of reducing the risks of earthquakes is achieved, the approach is like taking an axe to the problem when a scalpel would suffice.

The vast majority of places where the shale revolution has played out has experienced little or no problems related to earthquakes. In regions where oil and gas production is the backbone of the local economy, banning fracking to prevent earthquakes could be a huge financial blow to communities who depend on the oil and gas industry. So if banning fracking takes an axe to the earthquake problem, what might the scalpel look like?

WHAT TO DO ABOUT EARTHQUAKES

In the places where wastewater disposal has triggered earthquakes, state governments have taken a variety of approaches to reduce the risks, with varying degrees of success. (As I'll discuss more in the next chapter, state governments, rather than the federal government, do most of the regulating for the oil and gas industry.) Three states—Ohio, Kansas, and Oklahoma—provide useful examples of these different approaches.

One of the first high-profile quakes associated with the shale revolution took place in eastern Ohio, when a series of small earthquakes occurred near Youngstown in early 2011. These quakes were centered less than one mile from a relatively new wastewater injection well, and state inspectors began tests on the well roughly one month after the first quake was reported. By December of that year, the state had determined that this well, which injected fluids directly into the basement rock, was the likely cause of the quakes and ordered it shut.[18]

After some additional research by the state's Department of Natural Resources, the governor signed in July 2012 an emergency executive order allowing for temporary safeguards and the development of new guidelines for wastewater injection.[19] Although there have been a couple of relatively small earthquakes linked directly to fracking in Ohio,[20] the wastewater-induced tremors have essentially stopped, the state has implemented a high-tech monitoring program to minimize risks, and earthquakes are no longer a substantial issue in the state.

In Kansas, the state government acted more slowly than in Ohio. Kansas started experiencing quakes fairly regularly in December 2013 and established a task force to look into the issue in February 2014.[21] By September 2014, the task force produced a report detailing a plan to identify which injection wells were most likely responsible for the quakes. By April 2015, enough information had been gathered for the state to limit wastewater injection into the Arbuckle formation, the same formation that was at the heart of the problem next door in Oklahoma.[22] After seeing the number of quakes of magnitude 3.0 or greater grow from just two in 2013 to sixty in 2015, the injection limits helped reduce the number to just twenty-three in 2016.[23]

Just a few miles farther south in Oklahoma, policy makers moved slower still, despite the presence of more severe and widespread quakes. In 2011, the state experienced a quake registering 5.6 on the Richter scale, which the Oklahoma Geological Survey argued for several years was likely a naturally occurring event. At the same time, geologists from the U.S. Geological Survey and academic researchers were finding strong evidence suggesting that these quakes were, in fact, manmade. News reports suggested that the geologists in the Oklahoma office were not able to speak freely on the issue, perhaps not wanting to offend politicians or donors to the University of Oklahoma, which housed the Geological Survey and received substantial donations from oil- and gas-industry executives.[24]

As the number of quakes grew from 41 in 2010 to 579 in 2014, the state took no major action to limit wastewater injection. During these years, Oklahoma Geological Survey researchers and independent scientists were investigating the problem but did not come up with evidence strong enough to convince state policy makers to act. The Arbuckle had for decades been a major destination for disposed wastewater, but the increase in water volumes from the Mississippian Lime, Hunton, and other formations had substantially increased the pressures underground. This growth in pressure was then transmitted from the Arbuckle into deeper basement-rock formations, where the slippage occurred (as illustrated in figure 5.3).

Finally, in the summer of 2015, as the number of quakes continued to grow (and several months after Kansas had begun limiting injections into the Arbuckle), policy makers began taking action to reduce wastewater disposal. The state also beefed up monitoring efforts, required companies to show that their disposal wells were not affecting the basement rock, and took other actions, eventually resulting in detailed plans to reduce seismicity in key areas.[25]

As these plans went into effect, the number of quakes began declining. In 2016, Oklahoma experienced 623 quakes greater than magnitude 3.0, hardly a small number, but down from 903 in 2015. The pressures that had built up underground over the years were not going to disappear overnight. Like the Rocky Mountain Arsenal, where quakes continued for years after wastewater injection was halted, Oklahoma will be managing the wastewater issue for the foreseeable future. On September 3, 2016, a 5.8-magnitude quake, the largest in state history, struck northern Oklahoma, causing damage to historic structures, homes, and increasing the anxiety of the already unnerved residents of the region.[26] Like most other earthquakes that have occurred in the state in recent years, experts believe oilfield wastewater was the likely root cause.

While Oklahoma's quakes are likely to continue for some time regardless of the level of oil and gas activity, the risks are declining as the volume of wastewater disposed of underground decreases. In 2017, the U.S. Geological Survey released its annual seismic-hazard forecast, which provides estimates of the risk of damaging earthquakes across the country. While Oklahoma's levels were uncomfortably high, they were lower than the previous year's forecast, providing hope for continued improvement.[27]

The experiences of Ohio, Kansas, and Oklahoma illustrate three distinct approaches to preventing and managing the risks of wastewater-induced earthquakes. While these quakes are—when considering the thousands of disposal wells scattered across the United States—quite rare, the problems they cause can be severe.

Earthquakes associated with wastewater disposal aren't inevitable, and the issue has been blown out of proportion by antifracking opponents who warn that they will bring widespread damage. Now that there is a better understanding of the factors that contributed to Oklahoma's quakes, it should be relatively straightforward to manage the problem over time and reduce the risks to public safety and property damage for nearby residents. Other approaches, such as increased wastewater recycling, can also help by reducing the need for disposal deep underground.

However, Oklahoma's experience illustrates how, in some cases, policy makers may not be willing to act quickly to reduce the risks associated with oil and gas development. When the political and economic costs of intervening in the industry are high, governments have the difficult task of deciding how to weigh the risks and benefits of acting aggressively versus the risks and benefits of taking a slower approach. In the case of Oklahoma, a slower response prevented widespread disruption to the oil and gas sector but may have caused lasting damage to the public's accurate understanding of the connection between "fracking" and earthquakes.

SUMMING UP

In some instances, fracking has directly caused small earthquakes. But the quakes in Oklahoma that have made national news are not directly caused by fracking. Instead, they have been caused by the disposal underground of large amounts of wastewater generated from oil and gas wells. Underground wastewater disposal has taken place for decades, but if too much is pumped into the wrong place, earthquakes can be the result. These wastewater-induced quakes have occurred in parts of Kansas, Ohio, Oklahoma, and Texas and have in a few cases been large enough to cause damage to local property. While the scale of the quakes and the associated damage has been minor compared with large earthquakes in places such as California and Japan, wastewater-caused earthquakes need to be reduced. With good understanding and proper state-level regulation, the problem should be manageable, and earthquakes should be prevented. However, a limited amount of knowledge at the state level, coupled with a relatively slow policy response, particularly in Oklahoma, has led to years of quakes.

6

IS THERE ANY REGULATION ON FRACKING?

When I pulled up to the entrance of East Chapel Hill High School in North Carolina, on a typically humid spring afternoon, two competing sets of protestors were out in front. One group consisted of about fifteen men, mostly middle age or older, sporting matching red T-shirts and holding large white placards professionally printed with "Shale Yes!" (To get the joke, holler "Shale yes!" with your thickest Southern drawl.) Just across the driveway leading into the high school was a more colorful collection of twenty or thirty demonstrators, bearing a motley assortment of placards with slogans such as "We Can't Drink Money" and "Ban Fracking Now!"

In 2011, North Carolina state geologists identified a layer of shale about 3,000 feet below the central part of the state that held natural gas deposits: these might entice drillers to venture for the first time into the Tar Heel State. With shale development accelerating in other eastern states, including Arkansas, Pennsylvania, and West Virginia, many policy makers in North Carolina began to see visions of drilling rigs, rapid job growth, and increased tax revenues dancing like sugarplums on the horizon. Others, however, were more concerned about flaming faucets, heavy truck traffic, and the industrialization of the state's tranquil Piedmont region.

I was in Chapel Hill as part of my work, helping the state write a report on the potential impacts of shale development. The Chapel Hill meeting, where Shale Yes advocates faced off against antifracking protestors, was one of several opportunities for state regulators and the report's authors to receive public comments on a draft version of the report.

A panel of three high-ranking state officials serving under the state's Democratic governor, Beverly Perdue, sat atop the stage in the two-hundred-seat auditorium, listening quietly as each speaker made their case. The officials had clearly been instructed to show no favor and take no offense at the comments that were directed—or, more accurately—hurled in their direction.

Despite the fact that no fracking was occurring in North Carolina—and that no fracking was likely to occur given its modest shale resources—the vitriol and fear that the topic had engendered was on stark display. One speaker compared the report's authors to Germans in the 1930s and 1940s, idly standing by while Hitler's armies gathered strength, perpetrated genocide, and destroyed much of Europe (in this analogy, the spread of fracking represented the conquest by Hitler and his armies). Another speaker implored the audience to rise up against fracking, as it would lead to "ambiguous genitalia" among newborns.[1]

But the most common complaint focused on the notion that fracking was not regulated. Many commenters echoed a refrain that the 2005 Energy Policy Act, shepherded into being by Vice President Dick Cheney and a group of advisors with ties to the oil and gas industry, exempted fracking from the federal Safe Drinking Water Act, Clean Water Act, and other laws designed to protect the environment and human health.

Technically speaking, these comments were correct. Fracking, and oil and gas development more broadly, is exempted from certain portions of these laws. However, the commenters incorrectly assumed that since fracking was not regulated by these federal laws, it must not be regulated at all.

In reality, state governments apply extensive regulations to oil and gas development; require permits for various activities occurring before, during, and after drilling; and enforce those regulations. The key questions are, first, whether state governments, rather than the federal government, should be the ones doing the regulating; and, second, whether existing state regulations and enforcement are adequate to protect the environment and public health.

But before we try to answer those questions, let's lay the groundwork by understanding what is, and what isn't, regulated under the federal laws so often invoked by antifracking advocates.

FEDERAL REGULATIONS OF (AND EXEMPTIONS FOR) FRACKING

Historically, the federal government has not played much of a role in regulating the environmental issues related to oil and gas production.[2] In some specific cases, where oil and gas drilling occurs on land owned by the federal government or in the Gulf of Mexico, federal rules lead the way. But most shale-oil and gas development has taken place on private land (oil and gas production from federal lands has actually grown little since 2003),[3] where state regulations hold sway.

Nonetheless, given the intense focus on the federal role (or lack thereof) regarding shale development, it's worth going over some of the key federal statutes that exempt fracking and understanding the context for those exemptions. Leading the way in the discussion of fracking exemptions are the Safe Drinking Water Act, the Clean Water Act, and the Clean Air Act.[4]

The Safe Drinking Water Act, originally passed in 1974, is designed to protect drinking-water sources across the country, and it allows the EPA to set a variety of regulations to ensure drinking water meets certain standards. Many of these regulations, while set at the federal level, are enforced by state agencies (such as North Carolina's Department of Environmental Quality). Those state regulators establish detailed standards and enforce them to make sure public water systems (usually managed by local governments) are up to snuff.[5]

One important part of this regulatory scheme is the Underground Injection Control (UIC) program, which is designed to oversee wells used to store or dispose of a range of fluids and gases. As I described in the previous chapter, this program regulates the disposal of oil and gas wastewater, industrial and municipal wastes, and a variety of other disposal activities. Certain parts of oil and gas development are regulated under the "Class II" section of the EPA's UIC program.

Regulations apply to three types of wells: (1) wastewater injection wells; (2) enhanced recovery wells, which inject water, steam, or other fluids underground to increase oil and gas production; and (3) storage wells, which inject natural gas, oil, and other hydrocarbons underground so that they can be used later.

Hydraulic fracturing has never been regulated under the UIC program or any other federal program. Instead, the federal government has left it to the states to decide whether and how to regulate the fluids and explosives that companies regularly pump into wells to break up rocks and increase the flow of oil and gas.

However, this longstanding arrangement was called into question by a lawsuit a few decades ago, which for the first time raised the prospect of federal regulation of fracking. In this case, shale was not the issue. Instead, it was natural gas development from coalbed methane reservoirs (coal seams often contain substantial amounts of methane and are tapped in large quantities in parts of New Mexico and Colorado, among other regions). Fracking is used in some coalbed methane fields to increase the flow of gas from the coal seams.

In the 1990s, a nonprofit environmental law firm called the Legal Environmental Assistance Foundation, or LEAF, sued the EPA, attempting to force it to regulate fracking under the Safe Drinking Water Act. The case arose after complaints of water contamination from coalbed methane wells in Alabama's Black Warrior basin (the state environmental agency substantiated some, but not all, of these claims).[6] After a protracted legal battle, LEAF won the suit, and the state of Alabama was required to work with the EPA to develop a plan that would regulate fracking under the Safe Drinking Water Act.

The ruling, issued in 1997, applied only to Alabama, which is a relatively small producer of oil and gas, and did not result in federal regulation in any other states.[7] However, this case raised the possibility that similar lawsuits could arise elsewhere, eventually leading to federal regulation of fracking in places with more substantial production, such as Texas or Oklahoma. Preempting this potential outcome, the 2005 Energy Policy Act added language to the Safe Drinking Water Act ensuring that the LEAF case would not set a precedent that extended to other states. Because Dick Cheney led President George W. Bush's energy task force, which recommended this change, and because Cheney is the former CEO of Halliburton, which is hired by oil and gas companies to frack wells (and provide a variety of other services), this exemption of hydraulic fracturing came to be known as the "Halliburton Loophole."

However, when antifracking advocates raise the specter of Dick Cheney and his Halliburton Loophole in their explanation of how and why fracking has proliferated across the United States, they miss the point that the loophole effectively changed nothing. Oil and gas companies had used fracking for decades, regulated by states and not the federal government. They did not have to receive fracking permits from the EPA before 2005, and the Halliburton Loophole didn't change that. Of course, the legislation did make future lawsuits less likely and lowered the chances that the federal government would regulate fracking in the future, a prospect that most in the industry would welcome.

Oil and gas development is also exempt from portions of other federal regulations, including the Clean Water Act and the Clean Air Act. To a layperson, hearing that fracking, and oil and gas development more broadly, is exempt from the Clean Water Act, the next logical step would be to say, "Jeez. Sounds like oil and gas companies can dump whatever they want into the water." But as you probably expect by now, it's not that simple.

The Clean Water Act is one of the main regulatory tools by which the federal government prevents pollution to surface waters such as lakes and rivers. The act requires oil and gas operators, like everyone else, to receive approval before disposing of hazardous materials at the surface (whereas the Safe Drinking Water Act deals with *underground* drinking-water sources).[8] When advocates talk about exemptions from the Clean Water Act, they are usually referring to this: oil and gas companies are exempt from the portion of the act that requires builders to seek a permit for water running off from their construction activities into streams or storm drains. While builders of homes, businesses, factories, and other facilities typically need to meet certain standards to limit runoff, oil and gas wells are not subject to the same requirements.

This exemption means that drillers don't have to get permits from the EPA limiting the rainwater that runs off their well pads. Researchers have found that this exemption has had measurable consequences. Dirt and other sediment running off from construction sites cause environmental problems and economic damage around the country. In fact, a

2013 study found that increased sediment in some Pennsylvania rivers was attributable to increased runoff from the oil and gas well pads sprouting up to drill into the Marcellus.[9]

One tool that the federal government (during the Obama administration) had looked to use in regulating oil and gas companies was the Clean Air Act. Historically, oil and gas wells have not been tightly regulated under the Clean Air Act, but the Obama administration proposed new regulations to reduce the amount of methane and volatile organic compounds emitted from wells and other infrastructure (President Trump's administration has signaled that it will seek to halt and roll back these efforts).[10] As I describe in other chapters, methane emissions are an important contributor to climate change, and volatile organic compounds can pose health risks for those who work or live near well sites.

STATE REGULATIONS ON FRACKING

State governments, not the federal government, take the lead on regulating oil and gas production. These regulations can vary widely from state to state, and comparing one to the next, as some researchers have done, is a massive undertaking. I'll try here to provide a sense of what some of the key regulations are and describe how they vary between states.

But before diving into the details of some of these rules, it's important to point out that the mere presence of regulations on the books does not ensure environmental protection. Regulations may be designed or written poorly. In other cases, regulations simply don't follow best practices when it comes to protecting the environment or human health.

For example, we know that oil and gas drilling can cause methane migration, or "stray gas" (see chapter 3), potentially affecting groundwater sources for people living nearby. Some states require oil and gas companies to test nearby sources of water before, during, and after drilling, which can help determine whether drilling, rather than some other cause, is affecting drinking-water sources. These provisions provide some of the best information we can get about whether any particular water well has been affected by oil and gas development. But because they are costly, regulations mandating such testing are relatively rare and are not required in some major producing states, including Texas, Oklahoma, and North Dakota (discussed in more detail subsequently).

Another key issue is enforcement of regulations. A state government may have great laws on the books to protect the environment and public health, but if no one is checking to make sure the rules are being followed, they may as well not exist. Enforcement can also depend on the upper levels of management in state regulatory bodies. All regulatory agencies have discretion about how vigorously to pursue violations or how strictly to enforce penalties for companies that violate the rules. If their discretion tends to let polluters off the hook with slaps on the wrist, strong and well-written regulations may be meaningless in terms of real-world effect. On the other side of the coin, it is also possible that overly aggressive regulators could place undue burdens on companies.

Unfortunately for those of us who try to understand which regulatory systems work best, it's notoriously difficult to measure the strength and quality of enforcement, leaving us somewhat in the dark over the quality of environmental regulation at the state level. In the absence of good data on enforcement, there have been a number of efforts to document and quantify the strength of regulations on various state books. These projects, which require herculean feats of research and organization, help shed light on the key issues. In particular, a research project by the Washington-based think tank Resources for the Future (RFF) pulled together regulations from all of the major oil- and gas-producing states (excluding Alaska) to compare and contrast the state of shale-gas regulations.[11] (I started working at RFF in late 2016, after this report was released.)

To provide a sense of how these regulations work and how complex they can be, let's look at five examples pulled from the RFF report. Each of these issues is relevant to protecting human health and the environment, and they vary widely from state to state. Trying to describe all of the major regulations that appear in states would require a volume unto itself, but this far-from-exhaustive list is a good foundation for understanding some of the key regulatory decisions made by states.

"Setback" Requirements

Most states require oil and gas wells to be located a certain distance from homes, businesses, schools, rivers, lakes, and other potentially sensitive locations. The distances between a well and these locations are called "setbacks." Setbacks are applied to all sorts of developments (for example, most cities have requirements regulating the distance between a home or business and a public roadway or property line), and each state establishes setbacks from different types of structures and natural features.

When it comes to oil and gas development, the purpose of a setback is to put some distance between harmful accidents (such as well blowouts), releases (such as an oil or chemical spill), or emissions (such as volatile organic compounds) and the people, environment, or property they could harm. In some states, setbacks from buildings, including homes and businesses, are modest; 300 feet in Texas, for example. Other states, such as Colorado, Pennsylvania, and West Virginia, have setbacks of 500 to 625 feet.

In some regions, such as along Colorado's Front Range, rapidly growing communities sometimes encroach into territory where oil and gas drilling has gone on for decades. In these cases, state-mandated setbacks do not apply to the builders of new homes. In other words, new wells may not come too close to existing homes, but *new homes* can come as close as they'd like to existing oil and gas wells.[12] As a result, well sites pop up regularly in the backyards of new Front Range subdevelopments.

Some states also require setbacks from water sources such as rivers, lakes, and private water wells. For example, Pennsylvania requires that wells not be drilled within one thousand feet of a public water supply. In Texas, however, the RFF review found no setback requirements between water sources and wells, storage tanks, or other oil- and gas-production equipment.

Because the potential connections between oil and gas development and risks to human health are uncertain (see chapter 4), it's hard to know precisely how much distance is needed to prevent substantial public-health risks. In addition, it's not exactly clear how far a water source needs to be from a well to avoid all risk of methane migration. Because of this uncertainty, setback requirements are often something of a best guess on the part of policy makers. Of course, these decisions may also be affected by pressure from the energy industry, which typically pushes for shorter setbacks, and environmental lobbying groups, which typically argue for greater distances.

As in every other area of science-based policy, laws would ideally reflect the best research available. However, this ideal is rarely, if ever, achieved by any government. Nonetheless, as research gives us better information about the risks to human health and the environment, setbacks ought to be revised accordingly.

Water-Testing Requirements

As I have already noted, some states require oil and gas developers to test nearby water sources before, during, and after drilling and fracking. Typically, water samples are taken by an independent laboratory several weeks before drilling and tested to establish a baseline of water quality. Later, one or two weeks after drilling and fracking, that same lab will sample and test the water again. A few months further on, the process repeats once more (the exact time frames vary).

For states that require predrilling testing, such as Colorado, Pennsylvania, and Ohio, water wells within a certain radius of a new oil or gas well (ranging from 300 to 1,500 feet) must be tested. If these water tests identify changes in water quality during the testing period, the state begins the process of determining whether oil and gas development is the cause of the change. This process can be lengthy and contentious, primarily because determining with certainty the cause of any particular case of contamination is extremely complex. To deal with this issue, a number of states, including Pennsylvania, use what's called "presumptive liability," meaning that oil and gas companies are presumed to be responsible for any change in water quality unless they can show otherwise. If no substantial change is identified, testing stops. However, if landowners detect a change in water quality after the sampling process has ended, they can file a complaint with the state, which starts their own review.

Importantly, most major oil- and gas-producing states do not require companies to perform predrilling tests. As of 2012, Texas, North Dakota,

Oklahoma, California, Louisiana, New Mexico, and Wyoming didn't require predrilling water sampling. Because methane migration and other risks to groundwater loom so large in the public mind, it may be somewhat surprising that these states haven't moved to include this regulation. However, most of these states have seen far more oil and gas drilling in recent decades than Pennsylvania or Ohio have, and it's possible that the public is used to the types of risks and accidents that occur in the oilfield. It's also possible that the oil and gas industry, which would be required to pay for these testing regimes, uses its political clout in these states to make such regulations unlikely.

Another important issue for baseline water testing ties into the fact that oil and gas development in the shale era often occurs in more densely populated areas than many of the older oilfields in the sparsely populated West.[13] As a matter of common sense, wells drilled in more populated regions like the Marcellus or Utica are more likely to affect water supplies because of their proximity to the water wells serving nearby homeowners. In the Permian basin, Eagle Ford, or Bakken regions, where far fewer people live near oil and gas activity, it's easy to understand why predrilling water testing would be a lower priority.

Casing and Cementing Requirements

Because improper well construction can lead to stray gas and other types of pollution (see chapter 3), rules defining the casing and cementing of oil and gas wells are commonly the most detailed regulations applied to the industry. These rules cover how many layers of steel pipe, called casing, must be used to protect groundwater. Typically, companies must lay several layers of steel casing near the surface to seal off the well from surrounding soil and rocks. Most state regulations also cover how deep those different layers of steel piping need to go.

An even more detailed set of regulations is typically applied to the cement used to encase the steel pipes. The object of cementing is twofold: to secure the layers of steel casing firmly into place and to seal off the space between the well and the surrounding rock formations (as discussed in chapter 3, this space is called the annulus), which may contain salty brines, oils, gases, or other things that you wouldn't want moving up to the surface or into groundwater.

Cement regulations usually prescribe how deep the cement must go, the type of cement to use, the tests that must be performed to ensure the cement can withstand certain levels of heat and pressure, and more. Many states require cement to extend up from the oil- and gas-bearing rock formation by 500 or 600 feet, and some states, including Pennsylvania, require that a cement seal lines the entire well. In addition, some states require oil and gas companies to test their cement and send the results to the state regulatory agency, which helps ensure that the cement has been installed properly.

Of course, these regulations are not foolproof, and contamination can still occur in their presence. But because methane migration and other risks to groundwater tend to result from failures in casing and cementing, it makes a great deal of sense that regulators pay close attention to this issue.

Fracking-Fluid Disclosure Requirements

Because of the concern over contamination from fracking chemicals, much discussion has focused on the need for companies to disclose which chemicals are used, and in what proportions. All states have some degree of protection for trade secrets, meaning that companies do not need to disclose the precise proportions of every chemical used in the process. But existing rules typically require companies to disclose every chemical they mix into their fracking fluid, even while allowing for the protection of trade secrets.

Many states require companies to disclose the chemicals and fluids they use on the website fracfocus.com, though the details of what exactly must be disclosed vary from state to state. Most states with large-scale oil and gas activity, including Texas, North Dakota, Pennsylvania, New Mexico, and most other major shale states, require some degree of disclosure. Some companies also disclose additional details about the composition of their fracking fluids voluntarily. In all states, companies are required by the federal government to keep safety data sheets, which provide detailed information on the types of chemicals being used in drilling and fracking, onsite in case there is an emergency.[14]

Wastewater Requirements

Because of the large volumes of wastewater generated by oil and gas development and the risks associated with handling it, state regulation of wastewater also gets lots of attention. Like other topics, states vary widely on how companies must collect, store, transport, and dispose of the water.

Some states require companies to store wastewater in enclosed tanks, usually located on the well pad. Others allow companies to store wastewater in nearby earthen pits. For states that allow pits, other requirements come into play, particularly how the pits must be lined (usually with some type of impermeable plastic, though a few states allow companies to store wastewater in unlined pits), and how full the pits are allowed to be (to prevent spills over the top). However, as discussed in chapter 3, there have been numerous cases where these pits have leaked or spilled, resulting in local pollution of soils and water.

In addition to how water is stored onsite, every state works to regulate how wastewater is disposed of. Most states, working with the EPA under its Underground Injection Control program, allow companies to inject wastewater into disposal wells, though some (Pennsylvania in particular) have relatively few wells because of unfavorable geology (UIC wells are allowed only in places where geological conditions prevent the wastewater from contaminating any drinking sources). As a result, much of Pennsylvania's oil and gas wastewater is recycled to be used in other frack jobs or trucked to neighboring Ohio, where the geology is better suited to underground disposal.

Some states allow wastewater to be applied directly to roads for deicing or dust control, and some states allow wastewater to be stored in open pits until it evaporates. I'm not aware of any detailed research on the effects of these methods of disposal, but given the salts and other compounds found in wastewater, both may have some real, if geographically limited, negative environmental impacts.

LOCAL-GOVERNMENT REGULATIONS ON FRACKING

Because of concerns over environmental and health risks, some communities have called for additional regulation at the local-government level. Most drilling takes place outside of cities, often in rural regions like the Bakken, Permian, or Eagle Ford. In these unincorporated places, counties are usually happy to leave regulation to the states, though there are some exceptions.[15] Most of the debate over local control has played out in cities, particularly in Colorado, Texas, and Pennsylvania, where some have sought to add regulations on top of existing state laws to meet the expectations of residents.

The most extreme, and widely covered, form of these efforts is an outright ban or moratorium on fracking. In Colorado, several cities along the Front Range have banned the practice at various points. These cities, including Broomfield, Fort Collins, and Longmont, saw their bans overturned by the Colorado Supreme Court in 2016.[16] (A number of governments in states where fracking is nonexistent and highly unlikely ever to take place, such as North Carolina and Vermont, have also banned the practice, with virtually no real-world effect.)

In Texas, the city of Denton, which sits atop the Barnett shale, banned fracking in a 2014 vote. This was a substantial victory for antifracking campaigners, given that Texas is the country's biggest producer of oil and gas. Two factors helped make Denton a likely place to enact a ban. First, Denton is a college town, and student renters generally don't stand to benefit financially from production. Second, unlike most other Texas college towns, Denton had experienced substantial drilling activity, with almost three hundred wells within the city limits. Nonetheless, an oil and gas behemoth like Texas wasn't going to let this result stand: the state legislature invalidated the local ban in 2015 by passing a law that essentially banned local fracking bans.[17]

Short of moratoriums and bans, local governments in some regions have fought to manage oil and gas development along the same lines as they manage other industrial activities. In particular, local governments have sought to apply local zoning rules to oil and gas operations. Most cities use zoning rules to exert control over which types of development can occur in which places. For example, a city may require a cement plant or steel mill to be built in a part of the city reserved for industrial operations. A city may require that no liquor stores be within one mile of public high schools. A city may prevent strip clubs from setting up shop in the middle of downtown (in fact, the oil boomtown of Williston, North Dakota, voted in early 2016 to "strip" the business licenses of such establishments in its downtown).

However, most states prevent cities from applying these types of regulations to oil and gas wells. The logic of this approach is primarily to prevent companies from needing to comply with hundreds of different regulations—a patchwork of rules that would make it more costly and less efficient to extract oil and gas. In addition, smaller local governments, sometimes staffed by just three or four people, might not have the capacity or expertise to develop thoughtful regulations.

Another reason for limiting cities' zoning powers is that oil and gas resources can't be moved. A steel mill or strip club can be built in a variety of locations, but oil or gas wells essentially have to be drilled atop the relevant rock formations. If a state wants drilling to help boost its economy, it has an interest in regulating that development according to its own standards, whether the drilling takes place in a city or anywhere else. Alongside the economic interests of the state, oil and gas companies exert influence on policy makers to help ensure they can gain access to the best resources, wherever they may be.

Despite its "ban on bans," Texas allows for more local-government control of drilling and fracking than most states. For example, Fort Worth (another Barnett shale city) requires permits for each new well and restricts where drilling may occur. Fort Worth's permits include some of the regulations typically seen at the state level: annual fees, setbacks, notification requirements, and more.[18]

But in most other states, where cities have much less control, local governments ask the question: why should oil and gas be treated differently from every other industry? Aside from the justifications I have mentioned, there's no clear answer to this question. As a result, lawsuits continue to play out between local and state governments in Colorado and Pennsylvania.

WHO SHOULD BE DOING THE REGULATING?

With a reasonable background on how regulations are laid out today, we can turn to one of the key questions surrounding fracking and oil and gas development more broadly: Who should be doing the regulating?

The first issue to consider is whether there should be a strong federal role in regulating the industry. For most state officials in southwestern states including Texas and Oklahoma, or in rural western states such as North Dakota and Wyoming, the answer is a resounding "No!" This perspective makes some sense. Conditions vary widely from state to state, and one set of federal regulations would be unlikely to address each state's needs. Consider regulations limiting how much fresh water may be withdrawn from rivers and streams for fracking operations (these rules exist in a number of states). Such requirements would naturally look different in Pennsylvania, a state with ample water resources, relative to California, an arid state. In another example, a rural state like Wyoming might have no trouble mandating a thousand-foot setback between new wells and homes or businesses. But in a more densely populated state like Pennsylvania, this type of regulation would make it impossible to drill across large swaths of land.

Some research has attempted to assess the stringency of different state regulations,[19] but trying to quantify the precise effects of these different approaches is highly uncertain at best.[20] As a result, we don't have a clear understanding of which state's regulations are "best," especially since conditions on the ground vary so widely from state to state. What's more, the regulations on the books matter only if they are enforced. Primarily because data are limited, it's hard to quantify the quality of enforcement in any, let alone every, state. For example, if a state environmental inspector visits a well site with a leaky wastewater pond but fails to identify the violation, the state government, researchers, and the public may never know.

Despite this lack of reliable data, there is good reason to expect that the quality of enforcement varies among states. In Texas, which has over one hundred years of experience regulating the oil and gas industry, there is a large pool of well-qualified people who understand the technologies at play. Whether the state regulations are stringent or lax, Texas has plenty of potential employees who can lend their expertise to enforce them. In a state like North Dakota, where the size and speed of the Bakken boom caught almost everyone by surprise, it would be harder to pull together a strong team of regulators, inspectors, and other experts to enforce regulations.

The inevitable differences in enforcement among states is one argument for federal-level regulation. A single agency with a single set of standards would probably have more success in applying a consistent level of environmental monitoring and enforcement. There are other issues where federal regulation might make sense, particularly for areas where there is uniformity across states. For these issues, such as air emissions that could affect human health, the federal government is probably in the best position (or, to put a finer point on it, has enough money) to support high-quality research to understand the issues, then work to craft regulations based on the best available science. As already noted, the Obama administration took some modest steps in this direction. In 2015, the EPA issued drafts of several new regulations limiting air emissions from oil and gas well sites.[21] These regulations wouldn't have covered the entire industry—just new oil and gas wells—but did represent a step in the direction of more federal control.

One of the major challenges with developing and implementing regulations at *any* level of government is that the rule-making process is always influenced by politics. If the EPA were in charge, presidents and their appointees, regardless of political persuasion, would be able to adjust regulations and enforcement to suit their ideology and objectives. As the presidential transition between Barack Obama and Donald Trump highlights, environmental and energy policies can swing substantially from administration to administration.

Today, with most regulation in the hands of the states, it is the governors and their appointees that make policy decisions, again informed by politics. In Pennsylvania, for example, the 2014 election of Governor Tom Wolf, a Democrat, led to a substantial shift in tone and a more aggressive regulatory attitude toward natural gas development in the Marcellus shale. Under the previous governor, Republican Tom Corbett, state officials tended to lean further toward industry.

Ultimately, the choice of which level of government should take the lead in regulating oil and gas development comes down to a matter of trust in different levels of government. If you trust federal government agencies like the EPA to do a better job writing and enforcing rules, then you'll likely think that they are best positioned to handle the task. If you trust state agencies because of their local knowledge and expertise, you'll probably think they are the ones for the job. Unfortunately, there's no simple answer as to which approach is best.

SUMMING UP

Fracking, like other aspects of oil and gas production, is primarily regulated by state governments. Some have raised concern about policies like the so-called Halliburton Loophole, which exempts most fracking operations from federal regulations under the Safe Drinking Water Act. However, when this "loophole" went into effect, it didn't change much of anything happening on the ground, since fracking has never been heavily regulated by the federal government. Instead, state governments (and in some cases local governments) enforce regulations and require permits for all sorts of oil- and gas-production activities. Crucially, the details of regulations and the strength of enforcement vary from state to state, which can lead to regulations that range from well designed and well enforced to poorly designed and unenforced. Unfortunately, there is little in the way of hard data that allows researchers to identify where regulations are best crafted and enforced.

7

IS FRACKING GOOD OR BAD FOR CLIMATE CHANGE?

Driving south from Wheeling toward New Martinsville in West Virginia, I follow State Route 2, which snakes south along the Ohio River. Through the trees on my right, I see rusting tugboats push and pull hulking containers across the muddy water. The containers are filled with coal so black it almost glistens in the sun. On my left are cracked and scarred mountains, the sides of which have been blown apart to make way for this road. Within the mountains I can see a few of the rich coal seams that for many years have been the economic lifeblood of this region. Thin smokestacks, striped red and white, anchored in the narrow space between the yawning river and the snaking road, come frequently into view.

I drive under a tubular blue chute, about six feet in diameter, which shuttles tons of coal from within one of the mountains to a loading dock along the river. The dock boasts huge mounds of coal, soon to be shipped to the power plants lining the Ohio River that provide electricity for much of Appalachia and the Midwest. Some of this coal will also be shipped to power plants on the eastern seaboard, and some will travel to Europe and other international ports, where it will be burned for electric power and other industrial purposes.

The coal-loading docks and power stations are rusty and scorched. They're dirty, and they seem to embody a previous century—windows into the Industrial Revolution and the development of Appalachia. Indeed, since shale development took off in the mid-2000s, natural gas has helped speed the early retirement of many coal-fired power plants. Companies began developing West Virginia's shale gas in earnest around

2008, with thousands of Marcellus wells popping up in the hollows and hills around towns such as Clarksburg, New Martinsville, and Moundsville.

The Marcellus shale has created a new industry, and signs of the transition to gas are apparent along Route 2. Instead of huge piles of coal, parts of the Ohio are now lined by precisely cleared land studded with small yellow flags, indicating freshly dug natural gas–transmission lines. Enormous natural gas–processing plants, which separate the raw natural gas that emerges from the well into its different components, appear on the horizon, capped by bright orange flares dancing more than a hundred feet above the muddy water. Giant steel tanks filled with natural gas liquids such as butane and propane shimmer in the light, and miles of gleaming silver pipelines snake, Escher-like, around them.

This transition from coal to natural gas along the Ohio River has created new opportunities but also threatens economic hardship for the people who live in the region. It helps tell the story of what role natural gas, and its displacement of coal-fired electric power, might play in the transition to a lower-carbon economy. And it helps us understand whether natural gas helps or hurts society's need to reduce the greenhouse-gas (GHG) emissions that lead to anthropogenic (i.e., human-caused) climate change.

CHALLENGING THE KING

"King" coal, thanks to its age-old status as a low-cost electricity provider, has played a central role in the industrialization of nations around the world; it is also a crucial part of the local economies where it's produced. But when it comes to climate change, coal is public enemy number one. When combusted for electric power, coal produces about twice as much carbon dioxide (CO_2) as does natural gas. (Oil-based fuels are uncommon today in U.S. power plants and fall between coal and gas in terms of CO_2 emissions.)

As fracking has driven natural gas prices down to historic lows over the past several years (see chapter 9), coal-fired power plants have struggled to compete. In 2005, coal provided about half of all the electricity

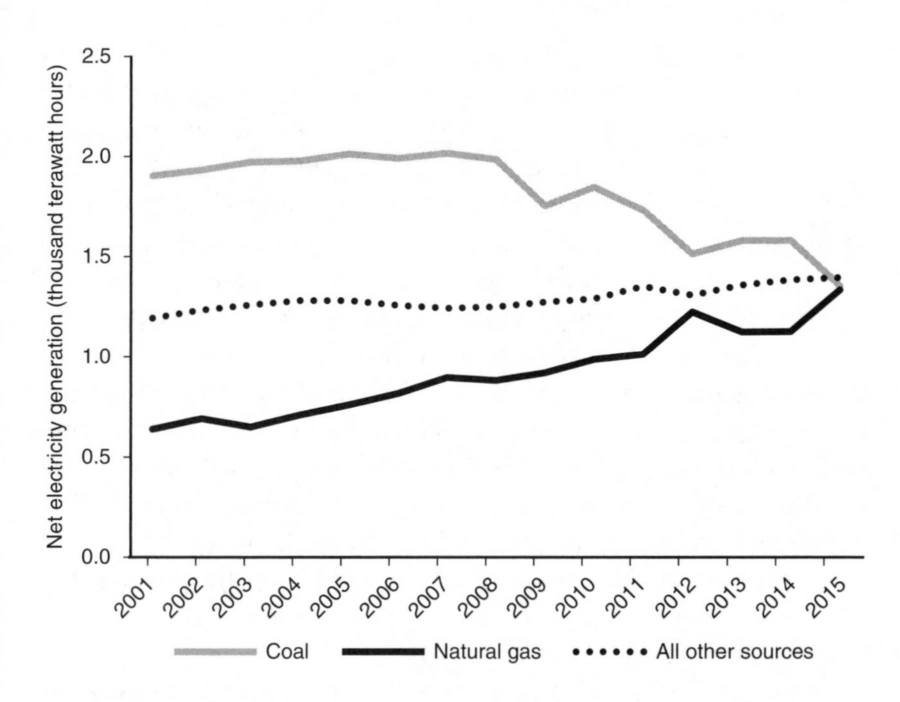

FIGURE 7.1 **U.S. net electricity generation by fuel**

Source: U.S. EIA, "Electricity Data Browser" (2017), https://www.eia.gov/electricity/data/browser/.

generated in the United States, while gas provided 19 percent. By 2015, coal and natural gas were essentially tied at 33 percent each. (Most of the remainder came from nuclear [20%], hydropower [6%], wind [5%], biomass [2%], and solar [1%]).[1] Perhaps more tellingly, the U.S. Energy Information Administration, which makes annual projections about future domestic and international energy use, projects that coal-fired power is unlikely to grow substantially in the future, while natural gas use will continue to climb.[2]

So more natural gas generally means less coal, which at first blush is good news for the climate. Emissions of carbon dioxide—the most prevalent greenhouse gas—will decrease, as will the emissions of mercury, sulfur dioxide, nitrous oxides, and the other pollutants that contribute to millions of premature deaths around the world each year.[3] We're already seeing the effect of this transition; in 2015, carbon dioxide emissions from the power sector in the United States were a little more than 20 percent below their 2005 levels, because of less coal, more natural gas, and more renewable sources, led by wind power. Proponents of fracking often cite the lower CO_2 emissions associated with natural gas compared with coal as an argument for the positive role natural gas plays in mitigating climate change.[4] However, the story isn't quite so simple.

First, while natural gas has pushed the U.S. market away from coal, some of the coal that now goes unburned in the United States is moving overseas, largely to Europe,[5] where it is being burned for power across the pond. Some have argued that coal exports have cancelled out the benefits of decreased coal consumption at home,[6] but that is not the case. Coal production in the United States has declined about 25 percent since its 2008 peak (coal imports have also declined),[7] meaning that the total emissions associated with U.S. coal have dropped, thanks to increased electricity production from natural gas and wind.[8] Nonetheless, the emissions associated with exported coal are not trivial and are rarely acknowledged by profracking advocates.

Second, natural gas also provides heating for homes and businesses (think of your gas-fired water heater, furnace, or stovetop), and it competes to provide these services with electricity and, to a lesser extent, fuel oil. About one-third of all natural gas consumed in the United States—about the same amount we use for electric power—is used for space and water heating in homes and businesses. Cheaper natural gas generally means more natural-gas-fired heating and fewer of the other options. This has a mixed effect on GHG emissions. Natural gas is cleaner than fuel oil, but it can be either more or less polluting than electricity depending on what types of fuels provide electricity in a given area. For example, natural gas will be less CO_2-intensive than electricity in a coal-heavy state like Missouri but more CO_2-intensive in a nuclear-, wind-, or hydro-heavy state like Washington or Oregon.[9] On average, natural gas–fired heating in the United States tends to be a little less CO_2-intensive than electricity, but as renewables generate more power in the future, heating with natural gas will become less climate friendly compared with electricity.

A third issue, the most contentious, relates to methane, a powerful greenhouse gas and the primary component of natural gas. A variety of academic studies conducted through the first decade of the 2000s found that greenhouse-gas emissions from across the natural gas system (including the methane emitted at well sites, methane leaked along the millions of miles of gas pipelines, leaks from other infrastructure, and CO_2 released during combustion) were easily better for the climate than coal.[10] However, much of this work relied on methane emissions estimates gathered by the EPA in the 1990s. When these data were called into question, so was the widely accepted view that natural gas was preferable to coal in the challenge of global climate change.

FIGHTING OVER METHANE "LEAKAGE"

While accounting for methane emissions from oil and gas wells, processing facilities, pipelines, and other infrastructure has been the province of oil and gas producers, regulators, and utilities for decades, the subject of methane "leakage"[11] has recently become a lightning rod in the debate over fracking.

Because it breaks down relatively quickly in the atmosphere, the contribution to climate change of each methane molecule varies over time when compared with CO_2, a more stable gas. Methane is 72 to 87 times more powerful a greenhouse gas than CO_2 over a twenty-year time frame and 21 to 36 times more powerful over a hundred-year time frame (these ranges exist because the climate system is—to put it mildly—complicated, making it difficult to estimate precisely the impact of methane. Estimates vary depending on assumptions made by researchers and have changed over the years).[12] The U.S. EPA has revised its estimates for oil- and gas-caused methane emissions over the past decade, roughly estimating that 1.3 to 2.0 percent of domestic natural gas production was being released as methane.

If roughly 4 percent of gas production was released as methane, the climate impact of natural gas would be roughly equal to that of coal over the shorter twenty-year window. If we focus instead on a hundred-year time frame, the leakage rate would have to be far higher—around 8 percent—for natural gas to do as much climate damage as coal.[13]

Assuming high methane emissions, natural gas starts to look like a GHG loser for heating, transportation, and other purposes.

As domestic natural gas production began to grow rapidly in 2007, so too came increased scrutiny of methane leakage and its climate implications. These issues were elevated to the front of the discussion over fracking by a 2011 paper from the researchers Robert Howarth, Renee Santoro, and Anthony Ingraffea.[14] In it, they argued that sky-high methane emissions from natural gas wells and pipelines (they claimed that up to 8 percent of gas produced from shale formations was being emitted as methane) actually made natural gas *worse* than coal for power generation.

The media coverage of the Howarth paper, especially from left-leaning outlets, was breathless and hair-raising. Alongside the robust list of reasons to oppose fracking, commentators could now add the claim that fracking was even worse than coal for the climate. For some advocates, natural gas—and fracking—had displaced coal as public enemy number one in the fight against climate change. As the energy analyst Michael Levi noted in a blog post, a major 2014 march against global warming in New York City felt like "an anti-fracking march that also happened to be about climate change."[15]

But oil and gas companies, as well as independent researchers, have pointed out that key assumptions in the paper are not supported by evidence or are demonstrably false. One example is the issue of "venting" natural gas. The Howarth paper assumes that 100 percent of the natural gas that returns to the surface during well drilling and completion is vented, that is, allowed to waft into the atmosphere as methane, inflicting large GHG effects. In fact, oil and gas companies "flare" most natural gas that returns to the surface during completion and early phases of production. Picture a bright orange flame flickering somewhere near the well pad: that's the operator flaring, or burning, the natural gas. When burned, methane is converted into CO_2, a less powerful greenhouse gas. At many well sites, companies flare very little gas and vent even less; instead, they capture most of the gas rising to the surface, put it in a pipeline, and sell it. Other problematic issues in the paper include the author's assumptions about leak rates from gas pipelines and their method of measuring the energy content of natural gas. A number of papers have pointed out these concerns, all of which biased Howarth et al.'s results toward making shale gas look worse for the climate than it otherwise would.[16]

Despite the paper's shortcomings, it sparked a rush of new research in the scientific community: Could the EPA really be that wrong about methane emissions? How bad is it? How do we fix it?

A COORDINATED EFFORT

Over the next several years, researchers from universities, think tanks, oil and gas companies, and environmental groups made a major push to get a better grasp of the scale of methane emissions from the oil and gas sector.

Perhaps the leading player in this research effort has been the Environmental Defense Fund, an environmental-advocacy nonprofit based in New York City. Following the old saw of "what gets measured gets managed," EDF led the way in coordinating a major push to measure methane emissions from the oil and gas supply chain in a variety of locations around the United States. Their effort was essentially an attempt to add new and better information to the EPA's estimates, which, as I have noted, relied on decades-old measurements that many believed were no longer representative of the quickly evolving industry.

To support its research, EDF coordinated millions of dollars from donors, foundations, and—controversially—oil and gas companies.[17] They recruited teams of researchers from universities and governmental research organizations to carry out the work, and they mandated high levels of rigor and an internal peer-review process. (According to news reports, they also required that researchers sign nondisclosure agreements promising not to reveal preliminary results, an unusual step in the world of academic research, indicating a desire for EDF to manage the press response to the research findings around such a controversial topic.)[18]

EDF also worked with oil and gas companies to help researchers gain access to well sites, compressor stations, processing facilities, pipeline networks, and other locations. A number of environmental-advocacy groups criticized EDF for working so closely with companies and expressed concern that this type of collaboration—along with industry funding of the research—would taint the findings.[19] But on the other hand, how do you research emissions from oil and gas equipment without gaining access to that equipment?

EDF has a long history of working with industry groups to research environmental issues, and this wasn't the first time they rubbed other environmental groups the wrong way on important issues. In the 1990s and 2000s, the group pushed for pricing carbon emissions through market-based mechanisms such as cap-and-trade, which was initially opposed by many environmental groups in favor of other policies to reduce greenhouse gases. Those alternative policies generally relied more heavily on governments dictating exactly how companies should reduce their emissions, rather than allowing market forces to determine the lowest-cost options for GHG reductions.[20] Over time, other environmental groups and governments around the world came to agree with the market-based approach, and cap-and-trade policies now operate in many regions, including the European Union, California, China, and more.

More substantively, any attempt to understand fully the extent of methane emissions from the oil and gas sector faces a number of challenges. First and foremost, the scale of the system is huge. There are easily more than one million active oil and gas wells in the United States and over a million more that no longer produce. In some parts of the country, such as northwestern Pennsylvania, where "Colonel" Drake kicked off commercial oil production, there are hundreds of thousands of abandoned wells lying dormant, unclaimed and unmapped, that could be emitting methane.[21]

But wells are just the beginning. There are more than two million miles of natural gas pipelines crisscrossing the United States (the distance between the Moon and the Earth is about 239,000 miles), and methane could be released from anywhere along those routes.[22] Additional leaks could come from the more than 1,400 compressor stations that push the gas through those pipelines and toward its final destination. There are almost four hundred natural gas–processing plants, like the ones I saw along the Ohio River, which together process more than 75 billion cubic feet of natural gas each day.[23] Still more leaks could come from any of the roughly four hundred natural gas–storage facilities, where gas is held in reserve for periods of strong demand.[24] Finally, methane can be leaked at any of the more than 1,700 power plants in the United States running on natural gas, or at any of the roughly 140 oil refineries, where natural gas is often used for distilling crude oil into gasoline, diesel, and other products.[25]

Given today's technology, precise measurement of every well, every compressor station, and every foot of pipeline is simply impossible (though new technologies such as high-resolution satellite-based methane sensors could provide this opportunity in the future). Instead, researchers need to measure emissions at certain locations, then make inferences about how representative those locations are of the larger system.

In addition, the inherent limitations of time and money mean that researchers typically collect data over periods of days or weeks rather than months or years. That wouldn't be a problem if natural gas systems emitted methane on a consistent basis from one week to the next. But as it turns out, emissions are often stochastic; that is, they can be unpredictable, and a compressor station that emits little or no methane for weeks or months may suddenly belch a large plume if a piece of equipment fails.

Despite these methodological challenges, a steady stream of research has emerged in recent years, much of it coordinated by EDF and even more produced independently by researchers from universities, think tanks, governments, and industry groups. This work has provided measurements that answer important questions yet raise others.

FROM THE BOTTOM, FROM THE TOP

Recent studies of methane emissions from U.S. oil and gas systems can generally be lumped into two categories: "bottom-up" and "top-down."

Bottom-up studies involve researchers setting up equipment at or near wells, pipelines, processing facilities, or compressor stations to measure methane emissions from those specific locations. Some have also equipped cars or vans with methane sensors, then driven through oil and gas fields or cities with dense pipeline networks.

Top-down studies, on the other hand, measure methane emissions across a broad swath of land, usually by flying specially equipped airplanes over oil- and gas-producing regions. In some cities, researchers have installed methane sensors atop office towers, allowing them to measure leaks from pipelines over a longer period of time.

While both types of studies—top down and bottom up—are valuable, they each have shortcomings. For bottom-up studies, the key limitation is the inability to measure large numbers of facilities. Methane-sensing equipment is expensive, making it hard to deploy at hundreds or thousands of locations simultaneously. As a result, even the most in-depth bottom-up studies typically report data for fewer than two hundred locations, making it challenging to draw nationwide conclusions.

For top-down studies, the primary challenge is attribution: determining the source of the methane picked up by the sensors. In the United States, the EPA estimates that in 2014, about 30 percent of manmade methane emissions came from oil and gas operations, whereas about 25 percent came from livestock: largely animal's digestive processes (cows, goats, sheep, and some other livestock essentially burp methane). Another 18 percent came from landfills, 10 percent from coal mining, and another 10 percent from manure.[26]

Because oil and gas production often occurs alongside these activities (for example, large oil and gas fields sit near cattle feedlots in Colorado, Oklahoma, and Texas and near coal mines in Pennsylvania, West Virginia, and Wyoming), it can be hard to tell which molecule of methane came from a gas well and which molecule is the aftermath of a cow's most recent meal. Researchers have different ways of dealing with this issue, but none are perfect. As a result, methane emissions estimates from the top down, while valuable, are rarely definitive.

Because of these complexities, the results of research into methane emissions are often confusing. Making matters worse, pro- and antifracking advocates have trumpeted findings that align with their goals, while seeking to undermine the credibility of results they find inconvenient. For example, some oil- and gas-industry advocates have questioned the motives of researchers who find high levels of methane emissions.[27] One article from *Energy in Depth* examining a study that identifies oil and gas sources as an important contributor to growing global methane concentrations concludes: "This report is just the latest example of the seemingly never-ending campaign to overstate (and subsequently overregulate) methane emissions from the oil and gas industry. Oddly, the report's data don't even support its hyperbolic narrative."[28]

On the other side of the debate, antifracking advocates, including Howarth himself, have jumped to highlight the subset of studies finding high emissions.[29] And when other studies emerge finding lower-than-expected emissions, some advocates have claimed that researchers were engaging in a cover-up. For example, an advocacy group called NC WARN, based in North Carolina, lodged a complaint against a researcher who has helped lead numerous studies on oil- and gas-based methane emissions. They argue that the researcher willfully distorted findings by obscuring the accuracy of methane-sensing equipment:

> Dr. David Allen, then-head of EPA's Science Advisory Board, has led an ongoing, three-year effort to cover up underreporting of the primary device, the Bacharach Hi-Flow Sampler, and a second device used to measure gas releases from equipment across the natural gas industry. Allen is also on the faculty of the University of Texas at Austin, where he has been funded by the oil and gas industries for years.[30]

WHAT THE RESEARCH SAYS

Dozens of studies have emerged in recent years examining methane emissions from different elements of the oil- and gas-supply chain.[31] And while there are exceptions, bottom-up studies have tended to find relatively low emissions, whereas top-down studies have estimated higher levels of methane leakage (remember, EPA's nationwide estimates show that roughly 1.3 to 2.0 percent of all natural gas is emitted as methane, and emissions would need to be above roughly 4 or 8 percent to be worse for the climate than coal over a twenty-year or hundred-year time frame, respectively).

A number of bottom-up studies that independently examined well pads, processing facilities, and compressor stations found lower emissions than the EPA's estimates.[32] If those studies were representative of the entire system, that would be good news indeed. But a few bottom-up studies have found higher emissions. A study carried out in Boston found that gas-distribution lines under city streets were leaking well above existing estimates.[33] Another study examining pipelines and processing plants in thirteen states found far more methane than the EPA had estimated.[34]

Just one recent (2015) analysis combines both bottom-up and top-down approaches, aggregating data from multiple studies of Texas's Barnett shale region. On the bright side, it found overall methane emissions to be roughly 1.5 percent of natural gas production. On the not-so-bright side, that was roughly 90 percent higher than the EPA had estimated for the region.[35]

But the most widely cited estimates of high methane emissions (aside from the Howarth paper) have emerged from several top-down studies of parts of Colorado, Utah, Los Angeles, and North Dakota.[36] These studies have estimated methane emissions ranging from 4 to 11 percent. Again, if those estimates were representative of the United States as a whole, they would seriously undermine the case that natural gas' displacement of coal has been beneficial for the climate.

But there are a few reasons to be cautious about extrapolating national conclusions from these regional studies. Along with the challenge of attributing emissions, which all top-down studies have to wrestle with, higher estimates of emissions have often come from regions with lower levels of natural gas production.

To illustrate why that matters, let's take a simplified example of two gas-producing regions: Parkerland and Monkville.[37] Parkerland is a natural gas powerhouse, with thousands of wells producing about 50 percent of the country's natural gas. Monkville is a smaller producer, supplying about 1 percent. Two studies are conducted looking at each region, and the results are startling: Parkerland has a leak rate of roughly 1 percent, and Monkville leaks 10 percent (see table 7.1). Monkville clearly has a problem, and it would be right to call for better control of its natural gas systems. But it would not be correct to translate the leak rates found in Monkville to the nation as a whole. In fact, the measurements made at Parkerland suggest that, on the whole, methane emissions are relatively low.

TABLE 7.1 Leakage rates of two gas-producing regions

	GAS PRODUCTION	LEAKAGE	LEAK RATE
Parkerland	50 units	0.5 units	1%
Monkville	1 unit	0.1 units	10%
Total	**51 units**	**0.6 units**	**1.2%**

TABLE 7.2 Emissions estimates of four natural gas–producing regions

	GAS PRODUCTION (BCF/D)	EST. EMISSIONS RATE	EST. EMISSIONS (BCF/D)	OVERALL EMISSIONS
Haynesville	6,593	1.5%	99	—
Fayetteville	2,797	1.9%	53	—
Marcellus	14,396	0.3%	43	—
Uintah	1,190	11%	131	—
Total	**24,977**	—	**326**	**1.3%**

Source: U.S. EIA Drilling Productivity Reports, for Haynesville and Marcellus; DI Desktop, for Fayetteville and Uintah basin.

Note: Wherever possible, I have obtained the data in this book from publicly available sources such as the EIA. However, in some cases, public estimates are not available for specific shale-gas plays. In those cases, I rely on DI Desktop, a proprietary data source, which allows users to search for production data from specific geological formations, basins, and other parameters.

Indeed, my hypothetical numbers roughly correspond to a set of studies conducted over multiple natural gas–producing regions over the past several years. In one study, researchers flew over three different shale plays: the Marcellus, Fayetteville, and Haynesville—all newer shale gas fields—to measure methane emissions.[38] They estimated a range of leakage rates between 0.2 percent and 2.8 percent for the various fields. In a study of the Uintah basin in eastern Utah, researchers using essentially the same techniques found leakage rates of up to 11 percent.[39]

So let's look at the numbers. If we take the midpoint estimates of methane leakage from the first study and the high leakage estimate from the second, we end up with a calculation similar to what we saw in Parkerland and Monkville (see table 7. 2). Because the Uintah basin produces a relatively small amount of natural gas, its leaks have less of an impact on national (and global) emissions relative to big producers such as the Haynesville and Marcellus. Of course, high rates of methane emissions in places like the Uintah basin are important and suggest that more needs to be done to control leaks in those regions. However, they do not confirm anything about nationwide emissions. And when it comes to evaluating the global climate impacts of natural gas, national—and, better yet, global—emissions rates are what we need to understand.

EMERGING CONCLUSIONS

While most studies have focused on emissions from specific regions, two recent papers have looked to evaluate the nation as a whole. One analysis aggregated findings from studies over the past twenty years; another utilized monitoring stations scattered across the country. Despite these different approaches, both came to similar conclusions: methane emissions from the oil and gas sector were likely in the range of 50 percent higher than the EPA's estimates.[40] If true, these results suggest that natural gas is not as climate friendly as we might have believed ten years ago but still easily better for the climate when compared to coal-fired power plants.

Along with these headline findings of leak rates, many of the studies discussed in this chapter reveal that substantial uncertainties still exist and that *better measurement is needed.* While EDF and the many researchers involved in tracking methane emissions have performed a valuable service in advancing the state of knowledge on the issue, it was never possible to answer all the important questions with a single round of studies, even though those studies were performed by high-quality analysts working with good intentions.

Another important finding is that a relatively small number of facilities appear to account for a large portion of methane emissions.[41] These so-called superemitters would presumably be the first place to target for emissions reductions. However, identifying these sites is not simple, given the millions of locations for potential leaks and the difficulty in predicting which locations are most likely to be leaky.[42] In some cases, such as the 2015 leak at California's Aliso Canyon natural gas–storage facility, the scale of the leak makes it easy to pinpoint.[43] But for smaller emitters, which make up the bulk of the problem, detecting leaks is neither easy nor cheap given today's technology. Responding to this challenge, the U.S. Department of Energy, EDF, and a variety of companies are investing in ways to spot methane leaks cheaply and quickly across the oil and gas system. If these technologies prove themselves, it should be relatively easy to plug the small number of superemitters, resulting in a large reduction of methane emissions at low cost.

A final point identified in the recent methane research is that newer systems tend to be less leaky than older systems.[44] This would imply that shale wells (which tend to be newer) leak less than conventional wells (which tend to be older). But methane emissions—whether from newly fracked wells in Pennsylvania or from eighty-year-old wells in Kansas—have the same climate impact no matter where they come from. Natural gas from an old well in Texas flows into the same potentially leaky pipeline as the gas from a brand-new shale well a few miles up the road. Because of the scale of the system, there is no substitute for good measurement.

And fixing these sites is just the first step. Recent research has focused mostly on the onshore U.S. oil and gas industry. We still have precious little information on how much methane is emitted from production offshore, such as in the Gulf of Mexico, and even less information about methane emissions in other parts of the world. For example, Russia is one of the world's largest producers of natural gas, and it maintains an enormous pipeline network that sends that gas thousands of miles to markets in Europe and Asia. How much methane is leaking from that system? Unfortunately, we don't have good data on this issue, and given the age of much of the Soviet-era infrastructure that moves the gas, there's reason to think the numbers may be higher than what we see in the United States.

THE EFFECT OF LOW PRICES ON ENERGY CONSUMPTION AND RENEWABLES

Most of the debate over the climate implications of the shale revolution has focused on the topics already discussed in this chapter: coal versus gas, and the extent of methane emissions from the oil and gas sector. But there's another important effect that often goes unmentioned: changes in energy consumption resulting from the fall in energy prices, which was brought about by the shale revolution.

Natural gas and, more recently, oil prices have dropped dramatically relative to their levels through most of the 2000s, and this is in large part because of new supplies brought onto the market from shale. These lower prices are good for U.S. consumers (we'll look more at the economic impacts of fracking in chapter 9). But low prices don't just put more pennies in people's pockets. Low prices also encourage them to consume more energy.

Consider the connection between gasoline prices and vehicle purchases. When gasoline prices are high, people buy more energy-efficient cars. When prices fall, they tend to buy more trucks and SUVs. This may not be the most far-sighted strategy when it comes to long-lived purchases, but it's what lots of people do.[45]

Along the same lines, lower heating and electricity bills mean more energy consumption. And when people use more energy, CO_2 emissions rise. The exact level of emissions growth that can be attributed to lower prices is difficult to know, since we cannot compare today's reality with the alternate reality where the shale revolution never happened. Suffice it to say that declining prices over the past decade have likely increased CO_2 emissions above what they would have been with higher prices. What's more, most projections show that energy prices are likely to stay relatively low in the coming years.[46]

While most of the debate over fracking and climate change centers around natural gas, fracking has also led to greater oil production. Driven by shale and other tight formations, U.S. crude production increased from 5 million barrels per day (mb/d) in 2008 to more than 9 mb/d by early 2015. This surge of U.S. production was a large factor—perhaps the leading factor—of the oil-price crash beginning in 2014, when prices fell from more than $100 per barrel to about $30 per barrel in early 2016.[47] These lower crude prices quickly resulted in a downturn in the efficiency of vehicles purchased by U.S. consumers[48] and probably global consumers as well.

Taking into account these fundamental responses to lower prices, the big question is this: Do the increased emissions caused by lower prices cancel out the reduced emissions from decreased coal-fired power generation? Answering that question requires considering yet another source of energy: renewables. Since 2010, the vast majority of new U.S. electricity-generating capacity has come from natural gas, wind, and solar. Looking forward, the U.S. Energy Information Administration estimates that of all the new power plant capacity that will come online in the next few years, about half will be fueled by gas, and a little less than half will come from renewables (primarily wind and solar).[49]

The emergence of wind and solar power in the United States and globally has been striking, and the two have grown far more rapidly than most experts had expected. But despite this rapid growth, natural gas appears to be edging out renewables in the near-term race. Because gas and renewables are easily the largest sources of new electricity in the United States, it is clear they are competing with each other for investment dollars.

Imagine this: an electric utility is retiring a sixty-year-old coal-fired power plant and needs to decide what type of plant will replace it. Without diving into the many details involved in making this type of decision, let's simplify and say that the utility's three best options are natural gas, wind, or some combination of the two.

In this scenario, natural gas and wind are competing head to head. Low natural gas prices will encourage the utility to build the gas plant and discourage them from building the wind. In reality, many power companies have faced this exact decision, as wind and gas are, in most regions of the country, the two cheapest options for new electricity.[50] So the key question is not *whether* cheap natural gas is hurting renewables but *how much* cheap gas is hurting renewables.[51]

It's not possible to answer this question easily, but let's start with some basics. First, a large part of the expansion of wind and solar in the past decade has been due to state-level programs called Renewables Portfolio Standards, or RPS. These programs, which have been implemented in dozens of states, require that utilities produce or purchase a certain percentage of their electricity from renewable sources. The wind and solar production that has been—and will be—installed because of these policies will not be directly affected by the low cost of natural gas.[52]

Another important consideration is whether and how low natural gas prices affect policy decisions. Numerous studies have demonstrated that, in the absence of policies that move the power sector (and the economy more broadly) toward a lower GHG-future, a low-cost supply of natural gas provides little climate benefit over the longer term, largely because lower prices induce people to use more energy, offsetting the emissions benefits of displacing coal.[53]

Under the Obama administration's Clean Power Plan, new regulations were adopted under the Clean Air Act that would reduce CO_2 emissions from power systems by some 30 percent below 2005 levels by 2030. These regulations, which have been challenged in court and which the Trump administration is working to dismantle, would help boost the competitiveness of wind and solar, along with continuing to encourage natural gas relative to existing coal plants. However, estimating the potential impact of policies like the Clean Power Plan and determining which source would benefit most—gas or renewables—is difficult to do precisely.

One useful estimate comes from the U.S. EIA's annual forecast,[54] which, among other scenarios, compares a world with the Clean Power Plan to a world without it. Under the Clean Power Plan, almost all new electricity generation from 2015 to 2030 comes from natural gas and renewables (primarily wind and solar), while coal declines. Without it, coal's share of the electricity mix stays roughly steady, at the expense of both gas and renewables.

As figure 7.2 shows, natural gas–fueled power generation with the Clean Power Plan grows to account for roughly 34 percent of total power generation in 2030 but remains roughly steady without the policy. Renewables grow under either scenario, but their rise is likely to be slower if the Clean Power Plan is removed. In 2030, wind and solar would account for 14 percent of the mix with the plan and 11 percent without. If gas and

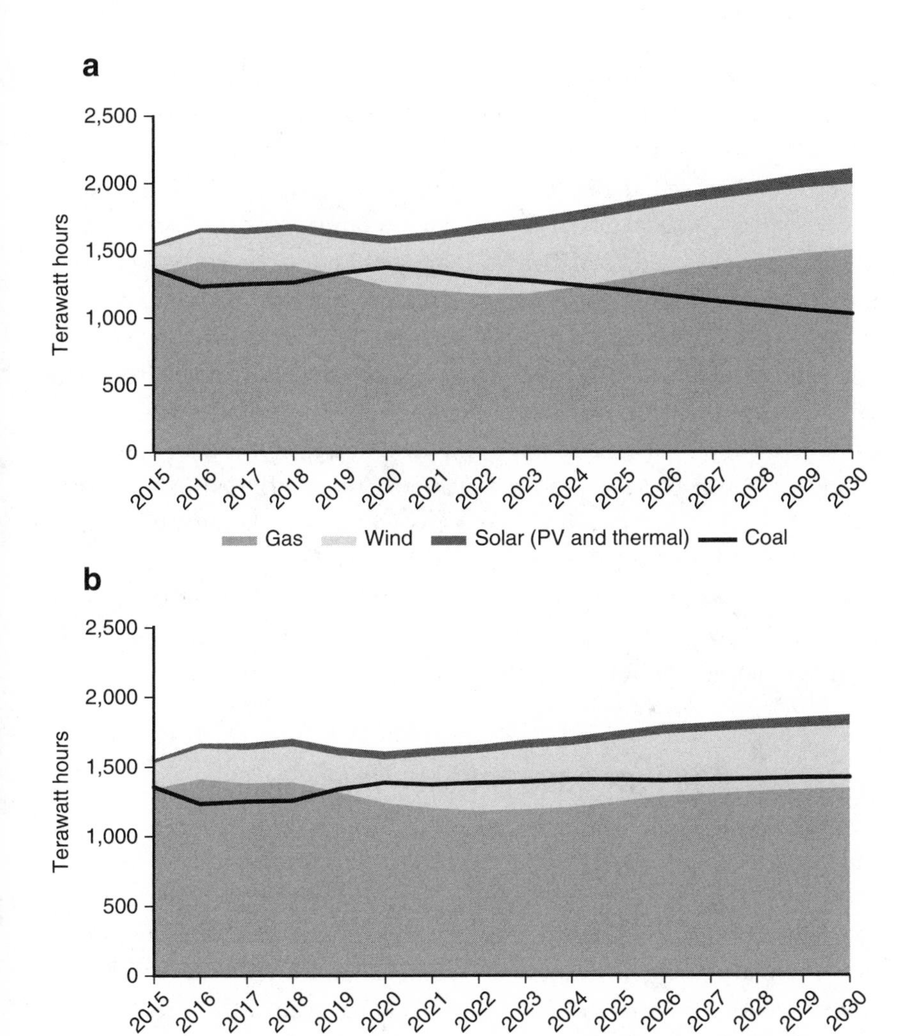

FIGURES 7.2A AND 7.2B **Projected U.S. electricity generation from select fuels with (a) and without (b) the Clean Power Plan**

Source: U.S. EIA, "Annual Energy Outlook" (2017), https://www.eia.gov/outlooks/aeo/.

renewables produce less electricity, the difference would be made up almost entirely by existing coal-fired plants (no new coal plants are projected to come online regardless of the Clean Power Plan).

So while natural gas is certainly not "killing" renewables, the basic point is clear: natural gas and renewables compete for investment dollars. Looking beyond 2050, the United States and the world will need to move toward a low- or zero-carbon economy to prevent climate change from causing large-scale damage to humanity.[55] That means that renewables and other low- or no-carbon sources such as nuclear and hydropower will need eventually to displace not just coal and oil but also natural gas (though technologies that remove and store CO_2 from the atmosphere or energy sector at large scales could support the continued use of fossil fuels).[56] However, most estimates of what will need to happen in our energy system to make this a reality involve natural gas as a substantial share of the global energy mix for decades to come.[57] As a result, it's important that we don't rule out natural gas but that we also don't let it stymie renewables or other technologies in the long run.

SO HOW DOES THIS ALL SHAKE OUT?

The long-term impacts of the shale revolution on climate change are complex, and a variety of forces will affect whether the increased oil and gas production it has allowed eventually moves us toward or away from a better climate future. Adding to this complexity are major uncertainties: the extent of domestic and global methane emissions, the effect of low natural gas prices on renewables and nuclear power, the effects of lower oil and gas prices on overall energy consumption, and policy decisions, particularly at the federal level.

But when viewed on a larger scale, fracking is probably neither a hero nor a savoir. Consider the worst-case and the best-case climate scenarios. In a worst-case scenario, where methane emissions are high and natural gas slows the deployment of renewables, nuclear, and other low-GHG technologies, our climate path would cause widespread damage to human well-being and the environment.[58] Economy-wide policies and coordinated international action would be needed to get the problem under control. But in the best-case scenario, where methane emissions are low and natural gas primarily hurts coal, the fundamentals of the situation remain unchanged: internationally coordinated action is necessary to deal with the problem.

Prior to the election of Donald Trump, probably the most important contribution of fracking to the fight against climate change stemmed from low domestic natural gas prices. These low prices didn't just reduce emissions by displacing coal; they also made it more politically and economically feasible for the Obama administration to move forward with the Clean Power Plan.[59] Without low-price natural gas, the costs of moving the power sector quickly away from coal would have been far greater.[60] While natural gas–fired electricity is projected to continue growing because of its abundance and low cost, its growth slows in a world without the Clean Power Plan, as coal reclaims some of its territory. Under the Trump administration, a world without the Clean Power Plan looks far more likely.

Outside of the Clean Power Plan, other steps can be taken today to reduce the greenhouse-gas impacts brought about by the shale revolution. As one example, the Obama administration in 2015 released a detailed plan on reducing methane emissions from the oil and gas sector. These limits—which some industry advocates have opposed and which are to be rescinded under the Trump administration—would have improved the GHG footprint of natural gas relative to coal and other fossil fuels. In 2016, the Obama administration released additional plans to limit methane emissions from oil and gas production on federal lands, though that has also been an early target of the Trump administration.

Despite the reversal of course at the federal level, some major natural gas–producing states are moving forward with methane regulations of their own. In particular, California, Colorado, and Pennsylvania have each announced their own reduction efforts. While these regulations will help improve the climate profile of natural gas, the effects will be limited unless adopted by other major producers such as Texas, Oklahoma, or Wyoming, where such action appears less likely.

So in the short run, fracking and cheap natural gas have been helpful in reducing greenhouse-gas emissions in the United States. Over the coming decades, policy decisions will play a key role in determining whether fracking slows or speeds a transition to a lower-GHG future. Taken together, the actions of the Trump administration to reverse course on methane and the Clean Power Plan will reduce the climate benefits of low-cost natural gas. While the shale revolution has provided a window of opportunity to enact more aggressive climate policies in the United States, it appears unlikely that the Trump administration will embrace such goals.

Still, fracking is not the key issue when it comes to dealing with climate change. A national (and, eventually, global) price on carbon, steadily increasing over time, coupled with investments in research and development for the energy technologies of the future, is the path identified by most researchers to tackle the global problem of climate change effectively and efficiently.[61] Increased oil and gas production from fracking hasn't changed that. And fracking bans wouldn't change that, either.

WHAT DOES THE OIL AND GAS INDUSTRY THINK ABOUT CLIMATE CHANGE?

As you'll recall, this book began with the tuba blasts of a polka band belting out tunes for bigwigs at CERAWeek, the enormous energy conference held each year in Houston, Texas. About six months after attending CERAWeek, I found myself listening to another band at another reception for another industry conference, in Pittsburgh, Pennsylvania. This time around, it wasn't a polka band but instead a classic-rock group covering 1970s and '80s hits by bands like Chicago and Earth, Wind & Fire. And when it came to differences between these two conferences, the musical selection for preconference entertainment was just the beginning.

For many casual observers of the oil and gas industry, it may seem as though the industry operates as a monolith. Indeed, one of the most well-worn phrases in political life is "Big Oil." And while there is some truth to the notion that different companies often strive toward the same policy and economic goals, there's far more diversity of opinion within the industry than many might expect.

One of the most important lessons that I've learned over several years of interacting with oil and gas companies around the country is that the industry is not a single behemoth but instead a network of individuals and corporations, each with their own attitudes and cultures. And while the culture of the oilfield is unlikely to be confused with that of academia, government, or most other industries, not all companies think alike. In fact, the differences from one industry conference to the next can be astonishing.

In Houston, where the presidents of Mexico and Rwanda, the secretary general of OPEC, the petroleum minister of Saudi Arabia, and executives from global leviathans including GE, Dow Chemical, and Siemens all made presentations, there was a decidedly international flavor. Instead of discussing topics from a U.S. perspective, panelists and presenters focused more on international energy issues, such as world oil markets, the increasingly global natural gas market, and geopolitics.

Perhaps surprisingly, one of the key themes running through the conference was climate policy. Even in sessions where it was not the headlining topic, climate change almost always came up. In discussion after discussion before hundreds of conference goers, Daniel Yergin (the founder of Cambridge Energy Research Associates, chair of CERAWeek, and author of the seminal history of oil, *The Prize*) made a point to ask a series of recurring questions on climate change: How will the 2015 Paris agreement affect your business? How might efforts to reduce greenhouse-gas emissions affect the global oil market over the next twenty years? How do climate policies affect the mix of fuels we use for power generation?

The questions were posed not as the topic of speculation but as a fact of life that had to be managed and planned for as a course of business strategy. No one asked "does burning fossil fuels lead to climate change?" or "are humans the main reason for increased GHG concentrations in the atmosphere?" but rather "how do climate change policies affect our business?" and "how can we reduce emissions from the energy sector?"

This attitude was in stark contrast to the one displayed several months later in Pittsburgh. Given the difference in political leanings between the states of Pennsylvania and Texas, one might expect a conference outside of the Lone Star State to show more openness when it comes to confronting the challenge of climate change. But that was not the case. From my first day at Pittsburgh's ShaleInsight conference, it became clear that climate change was not going to be on the agenda, at least not in any meaningful way. For an attendee who knew nothing about the topic going in, the three-day conference would likely leave the impression that climate change was some combination of four things: a hoax; not man-made; not worth worrying about; or, because of the rise of economies like China and India, a problem the United States can do little to address.

At the first evening's rooftop reception that welcomed attendees, the atmosphere was in some ways similar to what I had seen in Houston. In a lovely setting overlooking the confluence of the Allegheny, Monongahela, and Ohio Rivers, I scavenged from table to table assembling assortments of wine, beer, and hors d'oeuvres. I mingled with other guests, chatted about the latest in oilfield technology, and swayed to the tunes of Earth, Wind & Fire.

After a successful first round of gathering food and drink, I wandered over to a seating area where two large screens presented a new film: *Fractured: The Movie*. In the wake of *Gasland*, a number of proindustry films have attempted to counter *Gasland*'s key arguments and refute the elements of that film that have helped drive many to fear fracking. *Fractured* is one such film. Over the course of roughly ninety minutes, Mark Mathis, the director and star, sits in front of a blackboard and shares with viewers his perspective on how key terms in the energy debate (such as "fossil fuel," "clean energy," and "all of the above") are misleading and inappropriate.

One of the film's central points is that activists working to raise awareness about climate change are in it for the money. Leonardo DiCaprio, Al Gore, various environmental nonprofits, and others are—according to the film—predominately motivated by profit and are getting rich through the sale of unwarranted fears about climate change. Instead of engaging with the question of whether climate change is a problem and whether the public and policy makers should make any effort to reduce or avoid its negative impacts, the film instead attacks the messengers in classic ad hominem fashion.

The next morning, ShaleInsight began in earnest. Over the course of the day, executives from independent oil and gas companies, local politicians, and leaders from the Marcellus Shale Coalition spoke about key opportunities and goals for the Marcellus region over the coming years. Unlike CERAWeek, climate change was little more than a faint shimmer in the background of the day's presentations. When the topic did come up, it was invoked either as an afterthought or as a way to hammer the Obama administration's efforts to reduce CO_2 and methane emissions.

To begin ShaleInsight's final day, the conference welcomed a keynote speaker, Alex Epstein, an author and speaker whose most recent book is called *The Moral Case for Fossil Fuels.* As summarized on its website,[62] the book argues: "You've heard that our addiction to fossil fuels is destroying our planet and our lives. Yet by every measure of human well-being life has been getting better and better. This book explains why humanity's use of fossil fuels is actually a healthy, moral choice."

At a superficial level, there is logic to this argument. Access to affordable energy has enabled unprecedented improvements in human well-being, and fossil fuels have been the primary source of that energy. According to recent estimates, almost 3 billion people, primarily in sub-Saharan Africa, India, and southeastern Asia, lack access to affordable, reliable, and safe energy, which helps keep them trapped in poverty.[63] Bringing these billions out of poverty and providing them access to energy would be a moral triumph indeed.

But there are two reasons why these facts alone do not necessarily make fossil fuels a moral imperative. First, the argument ignores the reality of climate change, an enormous challenge made all the more daunting by the rapid expansion of fossil-fuel consumption around the world. Second, it conflates *fossil fuels* with *energy.* Access to *energy,* which may or may not be derived from fossil fuels, is the key to improving people's lives. And while coal, oil, and natural gas are in many cases the easiest and cheapest way to provide that energy, they are not a moral good in and of themselves. Indeed, the challenges of ensuring access to affordable energy while reducing the negative impacts of climate change are tightly intertwined, and both are fantastically complex. Talking about one without talking about the other hardly offers a "moral" path forward.

Finally, as if *Fractured: The Movie* and *The Moral Case for Fossil Fuels* were not sufficiently clear on the issue of climate change, ShaleInsight closed with a climate coup de grace: an appearance by the Republican presidential nominee (now president) Donald Trump. While Trump did not explicitly discuss climate change during his remarks (he focused instead on improving the prospects for coal, oil, and natural gas producers), his views on the topic are well documented on Twitter:

> The concept of global warming was created by and for the Chinese in order to make U.S. manufacturing non-competitive.
>
> (November 2012)

> Any and all weather events are used by the GLOBAL WARMING HOAXSTERS to justify higher taxes to save our planet! They don't believe it $$$$!
>
> (January 2014)

Wow.

Based on my experiences in Houston, Pittsburgh, and many other conferences and meetings, the oil and gas industry is not a monolith when it comes to views on climate change (and a suite of other issues). While the phrase "Big Oil" might be a common term, particularly among industry opponents, it rarely does justice to the range of goals, perspectives, and cultures that coexist within the industry itself. Looking forward, it seems likely that some big players in the industry will work toward constructive climate solutions, and others will delay an energy transition for as long as possible.

SUMMING UP

The shale revolution has reduced carbon dioxide emissions in the United States substantially over the past decade, primarily by providing a low-cost substitute for CO_2-intensive fuels, namely, coal. Looking forward, low-cost natural gas can continue reducing the level of overall greenhouse-gas emissions, but a couple of other things need to happen. First, governments need to adopt policies, ideally some form of carbon pricing, that encourage switching from GHG-intensive fuels (like coal) to lower-emitting fuels (like natural gas and—as the price of polluting rises—more heavily to renewables). The shale revolution has lowered the cost of natural gas, making it more economically and politically possible to begin this type of transition. Whether the United States takes that opportunity depends primarily on the actions (or inaction) of the federal government. Second, companies and governments need to make sure that natural gas wells, pipelines, and other infrastructure do not emit too much methane. As we approach the next century, society will need to reduce to a very low level the CO_2 emissions from all fossil fuels—including natural gas—if we want to prevent the worst effects of climate change.

8

WILL FRACKING MAKE THE UNITED STATES ENERGY INDEPENDENT?

Things were jumping at the Chili's restaurant near my hotel in Odessa, Texas. It was NCAA basketball tournament season, and I was out hoping to watch my Duke Blue Devils make their way to the next round. It was a Tuesday night, and although I knew that the region was booming, I didn't expect to have a hard time finding a seat at the bar in a strip-mall Chili's. Turned out I was wrong. I waited about fifteen minutes before a spot opened up, then nudged my way in to order a beer and have a seat to watch the game.

Odessa and Midland, together known as the "Petroplex," serve as the industry hub for the Permian basin, a vast area of some 86,000 square miles (a little bigger than the state of Utah) that includes more than forty counties in western Texas and eastern New Mexico. On my way to Odessa, through the seemingly endless West Texas scrubland, I had stumbled upon a monument commemorating one of the basin's first major wells: the Santa Rita well in Reagan County, which helped prove the potential for Permian oil in 1923. Since then, the Permian basin has produced massive volumes of hydrocarbons from a variety of geological formations layered like oil-soaked pancakes beneath the surface.[1] An oilfield that produces more than 10 billion barrels during its lifetime is known as a "supergiant," and while it consists of more than a single reservoir, the many productive layers of rock underneath the Permian have produced more than 30 billion barrels. And production is still climbing.

In the first part of the twentieth century, the Permian basin, along with other massive oilfields in Texas, Oklahoma, California, Louisiana, and elsewhere, made the United States the Saudi Arabia of its day. During that time, the United States produced far more oil than any other nation, serving as the chief supply source for itself and many of the industrialized economies of Europe. In fact, the ready supply of U.S. oil was an important factor in the Allies winning World War II, as the Axis powers needed to travel vast distances to secure petroleum, then transport those supplies back to where the war was being fought.[2]

The Railroad Commission of Texas, which was originally established to manage the growth of the state's rail industry, became the world's most powerful entity when it came to setting oil prices.[3] As Robert McNally describes in his recent book *Crude Volatility*, the Railroad Commission, in an attempt to keep global oil prices high enough to support continued industry stability, set quotas for producers across Texas.[4] If prices started to get too low, the Railroad Commission would restrict production, limiting supply and leading to an increase in prices.

But as oil production in Texas peaked and then began to decline in the 1970s, other production centers—in the USSR, Saudi Arabia, Iran, and elsewhere—surged to challenge the United States as the world's leading producer. More importantly, Texas was pumping at full capacity and had no dormant oil supplies to draw on in the case of surging demand—meaning that the Railroad Commission no longer had the ability to limit production to balance the global market.

In the commission's place, the Organization of Petroleum Exporting Countries (OPEC), led by Saudi Arabia and including Iran, Iraq, Kuwait, and Venezuela, took on the role of "swing producer," limiting production when prices were low (pushing prices upward) and increasing production when prices were high (pushing prices downward in an attempt to prevent oil demand from collapsing).[5]

The emergence of OPEC meant that oil consumers in the United States were now subject to the decisions of a group of leaders in far-off nations, some of whom expressed (and continue to express) hostility toward the United States and its international allies. When OPEC turned off the spigot, as it attempted to do in 1967 (with limited success) and again in 1973 (with more success), U.S. drivers felt pain at the pump. And because the United States had moved from being a net exporter of oil to a net importer, it and other import-dependent economies suffered from higher oil prices when OPEC limited production.

All of these factors came to a head in 1979 as spiking oil prices, long lines at gasoline stations (which were not in fact primarily caused by the OPEC embargo),[6] and dwindling domestic oil production led President Jimmy Carter to issue a series of speeches on energy, culminating in his infamous "Crisis of Confidence" speech from the Oval Office. Carter recited a series of quotes from recent meetings at Camp David, including the jarring statement: "Our neck is stretched over the fence, and OPEC has a knife." And, several minutes later: "Beginning this moment, this nation will never use more foreign oil than we did in 1977. Never. From now on, every new addition to our demand for energy will be met from our own production, and our own conservation. The generation-long growth in our dependence on foreign oil will be stopped dead in its tracks. Right now. And then reversed."

The United States did not achieve these goals. From 1977 to 2005, U.S. crude oil imports grew from 6.7 to 10.2 million barrels per day.[7] Nevertheless, the notion of energy independence—which focused on independence from imported oil—became a recurrent motif in the rhetoric of presidents for a generation to come.[8]

But what if we were to achieve energy independence? What if fracking enabled the United States, decades after Carter's speech, to reduce its oil imports to zero? What would that mean for prices at the pump, for our economy, and for our security?

CAN WE DRILL OUR WAY TO FOSSIL-FUEL FREEDOM?

Back in Odessa, the workers in dirt-stained Carhart overalls and mud-splattered boots joining me for beers and basketball were achieving something amazing. They were coaxing vast amounts of new oil from the Permian, which just a few years earlier many had thought was in permanent decline.

In the mid-1970s, production from the basin topped 2 million barrels per day (mb/d), falling to just 860,000 b/d by 2007.[9] But things had turned around. Led by production from rock formations with names like Spraberry (pronounced SPRAY-berry), Wolfcamp, and Bone Spring,[10] the Permian roared back to life, more than doubling its production over the next nine years to more than 2 mb/d in 2016. As production from other shale formations like the Bakken and the Eagle Ford boosted domestic supplies, U.S. crude oil imports fell from around 10 mb/d in the mid-2000s to a little over 7 million in 2016. News and opinion pieces wondered whether (and in some cases proclaimed that) the United States might become the "new Saudi Arabia"[11] and if energy independence might be within sight. To be sure, the United States was producing vast quantities of new oil, enough even to surpass Saudi Arabia as the world's top producer in 2014 and 2015.[12]

Now, imagine for the moment that this trend had continued—that U.S. producers further ramped up production to quench the entirety of U.S. demand for oil (roughly 19 mb/d in 2015). Would that have made us energy independent? The answer to this question hangs on how one defines "independent." If the definition of independence is to not import any crude, thereby sending $0.00 to foreigners for their oil, then producing all of the petroleum we consume would make us independent. But most understandings of "independence" imply that oil consumers would no longer be subject to the wide swings in gasoline prices that are broadcast every day on the giant digital price tags that festoon America's filling stations.

High prices, rather than the mere existence of imports, are the catalysts that cause drivers to moan and inspire politicians to pontificate. It was only in 2006, five years after he took office, that President George W. Bush declared America "addicted to oil." This declaration came on the heels of a rapid run up in prices to more than $50 per barrel, after having hovered for decades in the range of $15 to $20 per barrel.[13]

Figures 8.1a and 8.1b demonstrate that our interest in energy "independence" has a lot more to do with oil prices (and perhaps presidential elections) than it has to do with crude oil imports. Using Google Trends data since 2004, the number of Web surfers interested in energy independence spiked with oil prices in late 2008, then collapsed shortly thereafter as oil prices crashed. There is no such correlation between oil *imports* and interest in energy independence.

Our real interest in "independence" is not about driving oil imports to zero but instead about reducing the pain caused by unpredictable swings in prices. So the more pertinent question is: Will increased oil

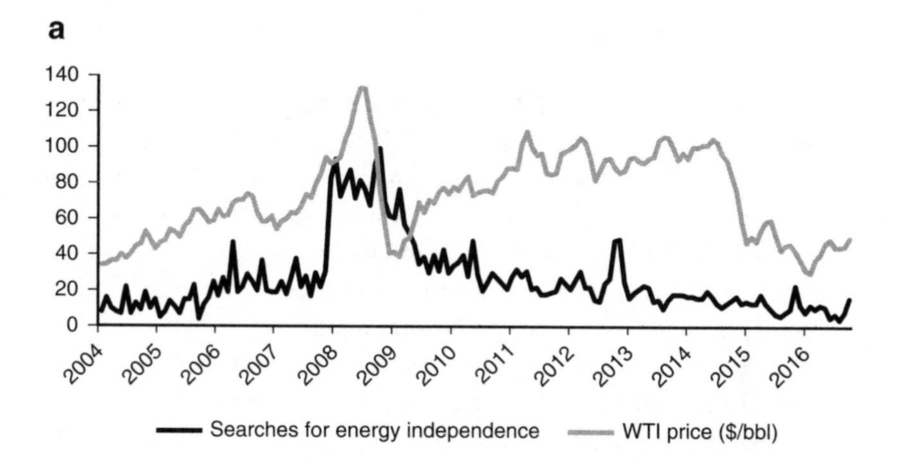

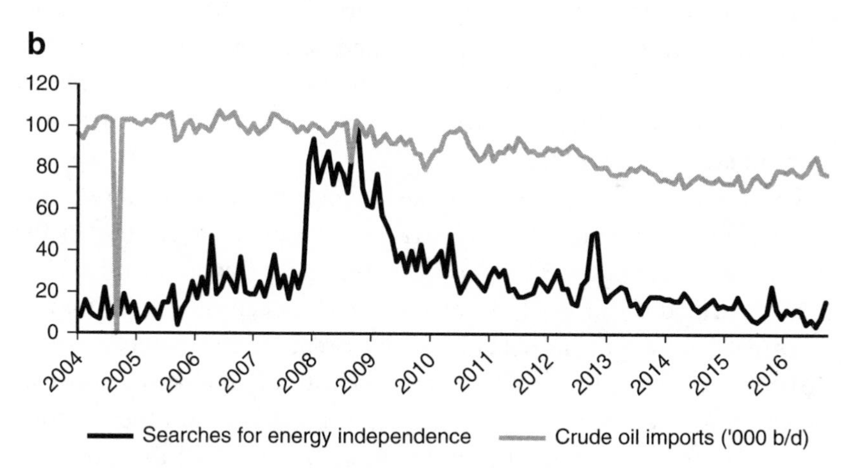

FIGURES 8.1A AND 8.1B **Interest in "energy independence": oil prices (a) and U.S. oil imports (b)**

Sources: U.S. EIA, "Petroleum and Other Liquids Prices," https://www.eia.gov/petroleum/data.php#prices; Google Trends. The Google Trends data were obtained by searching "energy independence" on https://trends.google.com/trends/. Results are not provided in raw number of searches. Instead, the results are normalized such that a value of "100" is the highest level of searches for the term and a value of "0" shows no searches for that term.

production in the United States make us independent of globally set oil prices?

The answer here is simple: No. And to illustrate this point, all we have to do is look north. Canada produced about 3.3 mb/d of oil in 2013. It exported 2.6 and imported 0.6 mb/d (Canadian production is centered in Alberta, most exports flow south to the United States, and most imports come through the East Coast), meaning its net exports were about 1.9 mb/d.[14] Canada is energy independent! And by a wide margin—they export more than three times as much as they consume.

So do Canadian drivers experience a warm glow when they pull into the gasoline station and pull out their wallets? No. In fact, Canadians—just like everyone else—pay the world market price for oil products. Because it's relatively cheap to move oil from one end of the globe to the other, and because oil markets are deep and liquid, both consumers and producers search internationally for the best prices. As a result, the oil market is global, and when crude prices rise or fall, they do so for every buyer in every corner of the world.[15]

When oil prices spiked in 2008 just before the Great Recession, the prices consumers paid in Canada tracked the prices next door in the United States. As figure 8.2 shows, U.S. oil prices (represented by West Texas Intermediate, or WTI) move in virtual lockstep with prices for two types of crude bought and sold in Canada, Brent Montreal and Canadian Heavy Hardisty. To be sure, there are modest differences in prices owing to the grade of crude oil, transportation costs, and other factors, but the central point remains: no one is independent of the global oil market.

Of course, the shale revolution *has* had an effect on oil prices. The surge in domestic production was a major factor in the global oil price collapse at the end of 2014, with continuing low prices through 2016.[16] And with sustained levels of production from shale, oil prices may continue to stay low, creating major economic benefits for U.S. consumers (and dragging down the profitability of producers). But that doesn't make the United States "independent." The prices paid by American drivers continue to be dictated by global market forces.

In some nations, especially developing economies (such as India or Thailand) and major oil exporters (such as Saudi Arabia and Venezuela),

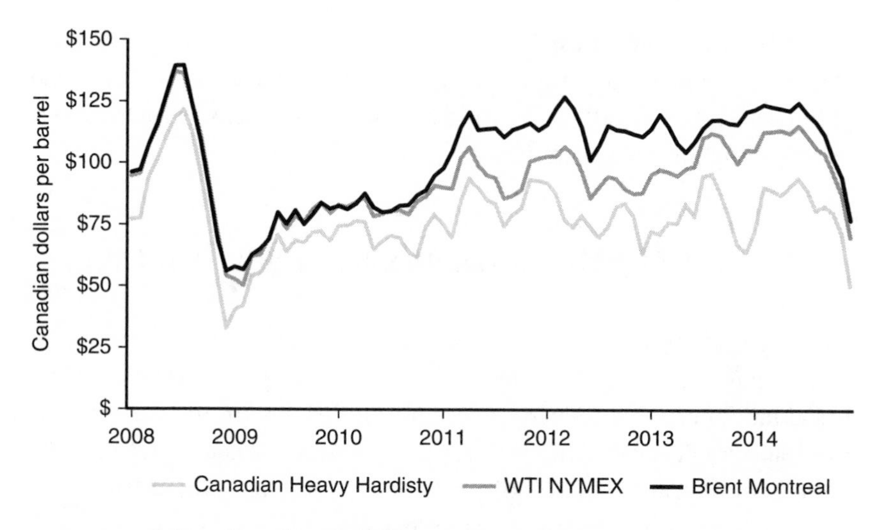

FIGURE 8.2 **Select Canadian and U.S. crude oil prices**

Sources: Canadian prices are from Natural Resources Canada (2015), "Oil Pricing," https://www.nrcan.gc.ca/energy/oil-sands/18087; U.S. prices are from the U.S. EIA (2016), "Petroleum and Other Liquids Prices," https://www.eia.gov/petroleum/data.php#prices.

governments subsidize oil consumption, resulting in lower prices for consumers. But that does not mean oil is worth less in those places. Instead, governments in these nations do one of two things. They either (1) purchase oil on the global market and then sell it to their citizens at a discount, costing taxpayers billions of dollars per year, or (2) they force the oil companies working in their nations (which are often government controlled) to sell their oil at a discount locally, subsidizing oil consumption at home while forgoing the higher prices that could be obtained by selling abroad.

These types of fossil-fuel subsidies are typically extremely damaging. Not only do they drain state coffers; they also starve government-controlled oil companies of revenue, preventing them from exploring for new fields or improving their technologies. What's more, the justification for these subsidies—that they help the poor who struggle to afford energy products—is generally swamped by the fact that the largest beneficiaries are the rich, who use most of the oil and thereby pocket the vast majority of the subsidies. Last but not least, fossil-fuel subsidies substantially increase pollution (including greenhouse gases and local pollutants) by artificially decreasing prices of CO_2-intensive fuel sources, encouraging greater consumption.[17]

THE ROLE OF NATURAL GAS IN ENERGY INDEPENDENCE

Although fracking has enabled dramatic increases in the production of oil *and* natural gas, this chapter has focused so far just on oil. Why? Political talk of energy independence typically centers around oil, which is the leading energy commodity imported by the United States and the world's most traded commodity. However, the United States has also been a net importer of natural gas for decades.

As with oil production, the United States led the world in gas production for much of the twentieth century.[18] But as consumption grew and domestic production declined, imports became an important part of domestic supply. In 1973, roughly 5 percent of U.S. natural gas consumption was imported—almost exclusively from Canada. But this figure grew steadily over the next several decades, and by 2005, imports accounted for 20 percent of U.S. consumption.[19]

With the United States already heavily dependent on oil imports, policy makers worried that the same story was about to unfold for natural gas. In fact, investors poured billions of dollars into liquefied natural gas (LNG) import terminals along the Gulf and East Coasts, anticipating a surge in needed supplies, with imports likely to come from global gas powers like Iran, Qatar, and Russia.[20] But as fracking helped unlock huge new stores of natural gas, many of these LNG-import terminals became unnecessary, and several are now undergoing costly conversions to export LNG instead.

Some commentators and investors have made the case for a dramatic expansion of natural gas consumption in the United States, arguing that increasing its use could lead to energy independence.[21] They argue that, because of this newfound domestic abundance, imported oil should be replaced with natural gas. Since oil is primarily used to fuel cars, trucks, buses, and airplanes, adopting this strategy would require converting millions upon millions of domestic vehicles to run on natural gas as well as retooling many of the more than 100,000 filling stations that line America's roadways.

Policy makers including James Baker III have long argued that natural gas, along with other alternatives, should be available to U.S. drivers in an effort to decrease reliance on oil. However, while there has been some increase in natural gas use in vehicles, a variety of issues, led by the costs of converting vehicles and refueling stations to natural gas, have prevented it from gaining much traction.

On the other side of the coin, although an increased use of natural gas in vehicles would reduce the United States' exposure to world oil markets, it would *increase* exposure to volatility in domestic natural gas markets. While natural gas prices have been low for several years, there is no guarantee they will stay there. Unlike oil, natural gas is relatively costly to transport and is mostly moved through pipelines, which tie together producers and consumers from specific regions. As a result, prices are typically determined by regional, rather than global, supply and demand.

Highlighting this issue, Jim Rogers, the former CEO of Duke Energy, gave a talk I attended where he quipped: "There are only three things in life that are certain: death, taxes, and volatile natural gas prices." And there is reason to think that natural gas prices are poised to rise over the long term, as demand for the stuff quickens. New power plants, chemical plants, and other industrial users are ramping up their demand for gas and would compete with motorists hoping to use it to fill their cars and trucks.

Although oil is the primary topic of concern when discussing energy "independence," natural gas is a fuel where dependence on foreign suppliers can have equally large geopolitical implications. Let's take an example, adapted and simplified from the 1973 Arab Oil Embargo. Imagine the United States finds itself in a geopolitical disagreement that leads Saudi Arabia and other Gulf nations to halt all oil exports to the United States, as occurred in 1973. Such a development would provoke concern over access to oil, and oil prices would undoubtedly rise. However, the physical impact would be minimal. Oil from the Gulf nations originally destined for the United States would find its way to another buyer—perhaps China, India, or Brazil. That shift in global oil trade would mean that oil produced in other nations—perhaps Russia, Nigeria, or Indonesia—would be freed up to travel to the United States. This type of shift would be disruptive and would create logistical challenges, but it would not lead to a physical shortage of oil in the United States.

Now let's imagine a similar geopolitical issue but this time use natural gas instead of oil as our fuel of choice. Again, I'll use an adapted and simplified version of real events. Let's say a political dispute emerges between Russia and Germany. Germany imports about 86 percent of its natural gas, much of it through pipelines from Russia. If Russia chose to enact an embargo, the Germans would have very little in the way of alternatives. Because natural gas is primarily moved through pipelines, the type of reshuffling that happens regularly in the oil market is simply not possible for natural gas. Germany could embark on massive new projects, building pipelines to other production centers, and it could build new LNG-import terminals, but completing these projects quickly would take several years at a minimum. In the meantime, how would Germans stay warm in the winter?

The lesson of these examples is simply that for natural gas, those without domestic supplies or supplies from friendly neighbors have reason to be concerned about depending on unstable, unreliable, or malevolent suppliers. But for the United States, the shale revolution, coupled with a good relationship with our northern neighbor, has prevented natural gas from becoming a new source of worry over energy independence.

WOULD IT BE GOOD TO BE ENERGY INDEPENDENT?

Why do we care about energy independence? One reason often cited relates to revenues flowing to unfriendly nations. In this theory, the United States helps support ideas and actions counter to its interests by purchasing oil from Saudi Arabia, Russia, Venezuela, or other nations where leaders or popular sentiment trends against U.S. objectives.

However, the world's largest exporters, such as the Middle East and Russia, ship most of their oil to Europe and Asia. American independence would do little to change that, especially given that Asian nations, led by China and India, will account for most of the growth in global oil demand in the coming decades.[22]

Instead, the key factor again relates to price. The decline in oil prices brought on in large part by shale development has had large negative effects on the economies of Russia, Venezuela, and other nations that rely heavily on high oil prices to fill public coffers and stifle dissent. In addition, new LNG exports from the United States may reduce the dominance of Russia on Europe's natural gas market, weakening Russia's negotiating position with its European neighbors.

But these are not arguments for independence. In fact, they are arguments for the opposite. If the goal of the United States is to help its allies and weaken its foes, the shale revolution enables that goal *only* if it enables deeper integration into global energy markets. This integration allows the United States to provide additional energy supplies to those currently on offer in the international marketplace, reducing the leverage of incumbent exporters. Indeed, even if the United States could isolate itself from the global oil market, this isolation would tend to enhance, rather than diminish, the influence of energy powerhouses such as Russia or Saudi Arabia.

There are also strong economic arguments for tying into global markets. To illustrate the value of integration, consider another product: martinis. If the United States moved from its relatively open approach to the global liquor trade and instituted an "independence" policy for gin (I prefer gin, rather than vodka), the price of martinis would shoot up overnight. While there are plenty of good gins crafted in the United States, my favorites tend to be British. The independence policy would reduce the volume of dollars sent across the Atlantic, and American gin production would increase. However, the years it would take for U.S. distillers to get their operations up and running would be a painful period, and it's unclear whether they would ever be able to match the quality of the Brits' gin. And what if—heaven forbid—the United States has a bad juniper season? Gin prices would surge even higher, quality would fall further, and national happiness would surely slump.

While oil and gin markets are very different, the example illustrates an important and basic idea familiar to anyone who's ever taken an economics course: in most cases, tying into global trade is a good idea. And while oil markets are much more geopolitically important and complex than the transatlantic gin trade, the pros of international trade for these goods easily outweigh the cons.

COULD WE LEAVE THE MIDDLE EAST?

Another common argument for energy "independence" in the political sphere is that achieving this goal would allow the United States to reduce its involvement in the messy politics of the Middle East. For decades, an uneasy alliance between the United States and Saudi Arabia has endured despite differing approaches on a variety of issues, including religious freedom, human rights, and terrorism.

For casual observers, it may seem obvious that decreased imports of Middle Eastern oil would translate into the ability to reduce American involvement in the region. And largely because of increased oil production at home, imports from the Persian Gulf have fallen from around 2.5 mb/d in the early 2000s to about 1.5 mb/d in 2015 (see figure 8.3).[23]

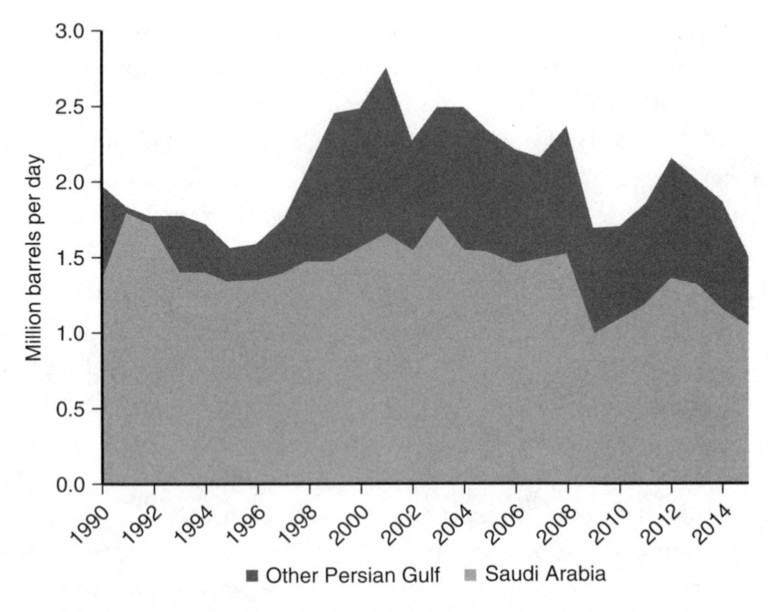

FIGURE 8.3 **U.S. oil imports from Persian Gulf nations**

Source: U.S. EIA (2015), "Petroleum and Other Liquids Imports/Exports and Movements," https://www.eia.gov/petroleum/data.php#imports. "Other Persian Gulf" consists of Bahrain, Iran, Iraq, Kuwait, Qatar, and the United Arab Emirates.

Another thought experiment here is helpful. Let's imagine that imports from the Persian Gulf fell to zero. What would happen to American involvement in the region? Most likely, very little would change. There are several reasons why. First, as the foreign affairs and energy expert Meghan O'Sullivan pointed out in a 2013 column, oil is not the only reason for the United States' interest in Middle East politics. The issues of terrorism, nuclear proliferation, human rights, and Israel will not go away based on how much oil and gas are pumped at home.[24] And while Saudi Arabia and other Gulf nations may not be ideal partners on human rights or terrorism, their powerful standing in the region means that the United States has a strong interest in working with them to make any progress in these complex areas.

The second key reason why American interests in the Middle East will continue to be strong relates to the global nature of oil markets. Regardless of what happens in the United States, the Middle East is—and will continue to be—the key facilitator of the global oil market. In 2015, the region accounted for 44 percent of global crude oil exports, far more than any other crude-exporting region (see table 8.1).

Since a disruption in the oil supply anywhere in the world can cause price spikes to all consumers, political instability in the Middle East

TABLE 8.1 Crude oil exports by region, 2015

REGION	CRUDE OIL EXPORTS (MB/D)	SHARE OF GLOBAL EXPORTS
Middle East	17.7	44%
Eurasia	6.7	17%
Africa	5.7	14%
North America	4.9	12%
Central and South America	3.5	9%
Asia and Oceania	1.0	3%
Europe	0.2	1%

Source: BP, *Statistical Review of World Energy*, 2016, http://www.bp.com/en/global/corporate/energy-economics/statistical-review-of-world-energy.html.

means higher prices for everyone. And since the United States is by far the world's largest oil consumer, it has an outsized interest in trying to maintain stable oil prices. Partly because of this interest, the U.S. Navy's Fifth Fleet patrols the region to secure the free flow of trade in and out of the Gulf. As Vice Admiral John Miller, the commander of the Fifth Fleet stated in a 2015 blog post:

> Although the United States is exporting more energy . . . we remain tied to the global economy that depends on the uninterrupted supply of oil and gas from the Middle East. This uninterrupted supply can be placed at risk due to rising political instability and regional conflict such as we now see in Iraq and Syria.
>
> In particular, Iran continues to develop an increasing capability to threaten commerce transiting the Strait of Hormuz. A disruption in energy supply would immediately and significantly affect the global economy.
>
> And that's where we come in.[25]

As much as some might like to untangle the United States from its complex and contentious relationships with the Gulf powers, the reality is that American interests will remain in the region for the foreseeable future, regardless of how much new oil and natural gas fracking helps unlock.

Looking forward, the only way to reduce dependence on global oil markets is to use less oil. Increased vehicle efficiency, higher fuel taxes, and—in the longer term—a transition to other fuels such as electricity or hydrogen can be a far more effective way to insulate American consumers from the price fluctuations of the global oil markets, not to mention the benefits of reducing greenhouse-gas emissions associated with oil-based transportation.[26]

SUMMING UP

The familiar trope of "energy independence" typically refers to U.S. oil imports, but the way it is used by politicians and in most media is often counterproductive. In fact, the shale revolution has *increased* the United

States' interdependence on global energy markets, and that's a good thing. Isolating the country from global energy supplies makes it more, not less, vulnerable to disruptions that can cause prices to spike. While fracking has reduced oil imports from the Middle East and other regions, consumers are still affected by prices that are shaped by events around the globe. With or without fracking, any shocks to oil production such as political instability in the Middle East would raise prices for everyone, even major oil-producing countries. For example, Canada is energy "independent": it's a major oil exporter. But Canadians feel the same pain that U.S. consumers feel when oil prices spike. More oil production in the United States means we are sending less money abroad, but more production at home does not make us "independent" from the global forces that shape the oil market.

9

IS FRACKING GOOD FOR THE ECONOMY?

By 2013, Rifle, Colorado, had slowed down quite a bit. Drilling took off there in the early 2000s, and in 2005 alone nearly eight hundred new wells were drilled in surrounding Garfield County. By 2008, drilling surged further, with over two thousand wells completed in the county. Rifle was booming. Other cities with names out of John Wayne films—Grand Junction, Parachute, Rulison, and Silt—were also feeling the effects, but Rifle was the epicenter.

Driving around Garfield County, which hugs the western slope of the Rocky Mountains, visitors are treated to postcard-worthy views at each turn. Snow-capped mountains loom just out of reach for much of the year, and multihued parched mesas stretch farther than the eye can see. Western Colorado, and Garfield County in particular, has been home to large-scale oil and gas development for decades, with the government and private companies seeking to coax its oil and natural gas out of the ground with a variety of unconventional methods.

Some methods have been more unconventional than others. In the 1960s, the Department of Energy sponsored a test to produce natural gas by detonating a nuclear device several thousand feet below the surface near Rulison, about ten minutes west of Rifle.[1] While the test produced large quantities of gas, there was a problem: the gas was radioactive.[2] According to a Department of Energy fact sheet, initial gas production showed "unacceptable levels of radioactivity," though radiation levels declined as more gas was produced.[3]

After other tests in neighboring Rio Blanco County (Project Rio Blanco) and New Mexico (Project Gasbuggy), the effort to produce natural gas using nuclear explosions was shelved. Despite the laudable goals of these tests—to develop beneficial, rather than destructive, uses of nuclear blasts—it's not hard to understand why consumers might be a little wary of radioactive gas being piped into their kitchens and basements.

The federal government and energy companies have also worked together for decades to explore the vast oil-shale reserves[4] that line the mountains along Interstate 70, which bisects Garfield County.[5] The potential oil resources locked inside these rocks are enormous, several times the magnitude of even the biggest shale plays, including the Bakken, Eagle Ford, or Wolfcamp. However, decades of experiments haven't led to profitable extraction, largely because it takes too much energy to coax the oil from the rocks.

But it was neither oil shale nor nuclear detonations that produced the most recent boom in Garfield County. Instead, it was the natural gas–bearing formations of tight sands scattered below Rifle and much of the rest of the region. "Tight sands" are more permeable than shale rocks but not as permeable as "conventional" reservoirs such as sandstone or limestone, where oil and gas flow more easily. This lack of permeability had made it difficult for companies to produce commercial quantities of gas from tight sands profitably.

However, high natural gas prices in the mid-2000s encouraged more companies to develop the sands in Garfield County and other regions, such as Wyoming's Green River basin and Utah's Uintah basin. Coupled with high gas prices, directional drilling and hydraulic fracturing helped make the plays profitable. As a result, dozens of operators moved into the region, and natural gas started flowing in enormous quantities. Workers streamed in to conduct geological studies, operate rigs and construction equipment, build pipelines and processing facilities, drive eighteen-wheelers to and from the well pads, and much more. New homes and developments sprouted rapidly in the shadow of the mountains, and while the oil and gas companies were not the only reason for rapid population growth (retirees and mountain lovers were also moving to the city, roughly a ninety-minute drive from Aspen), Rifle was bursting at the seams.[6]

But as 2008 drew to a close, the Great Recession made its presence felt in dramatic fashion. Not only did natural gas prices collapse, but the housing boom, which had been the other engine of regional growth, also ran out of steam. Drilling rigs started disappearing, followed by the workers. By 2013, when I visited for the first time, only a few rigs were drilling in Garfield County, and it was easy to find a seat at the local rib joint across the street from my hotel. McDonald's wasn't paying fifteen dollars an hour anymore, and the shelves at the new Wal-Mart were no longer stripped bare by rabid demand.

Rifle, and Garfield County more broadly, is an example of how oil and gas development can lead to a rapid and large-scale boom, sending local economies into hyperdrive and providing amazing job opportunities. But it also shows the risks: a booming economy built on volatile oil and gas prices cannot last forever, and planning for the future in the midst of a drilling frenzy can create an entirely different set of problems.

THE BIG PICTURE

In 1998, oil and gas development was a relatively small part of the U.S. economy. In total, the economic contribution of oil and gas extraction to gross domestic product (GDP) was about 0.4 percent. This category—oil and gas extraction—includes the core of the oil and gas business: anyone who works directly for companies that find and produce oil or natural gas.[7]

By 2014, at the peak of the shale boom, this sector's contribution to the economy had more than quadrupled, contributing 1.7 percent of GDP. An increase from 0.4 percent to 1.7 percent may not sound like a lot, but in raw numbers, 1.7 percent of GDP in 2014 represented $294 billion.[8] That's larger than the economy of the entire state of Arizona or the nation of Chile. And that $294 billion is from just those directly employed by companies focused on the extraction of the oil and gas itself. The numbers grow substantially when you start to include those that provide services to oil and gas companies, including well drilling, fracking, and more.

This 1.7 percent figure also does not account for indirect contributions to the economy. For example, a petroleum engineer working in Rifle buys dinner at the barbeque restaurant, then picks up a six pack of beer at the corner store. The employer pays for the engineer's hotel, or maybe the engineer rents a place in town. The family might buy a new TV or new car.

These indirect and induced economic effects are large but also subject to a wide range of uncertainty, making them difficult to calculate precisely. Because researchers can't track every oil and gas worker, estimating their indirect and induced effects on the economy means that economists have to make assumptions (basically informed guesses) about all sorts of things: How much of the engineer's income is saved, and how much is spent? Where does the engineer spend it? What does the engineer spend it on?

As a result, estimates about the overall economic contribution of any industry, including the oil and gas industry, can be hard to pin down. Despite the challenges, researchers—some independent, others supported by industry groups—have produced dozens of studies over the past five to ten years that estimate the local economic and employment effects of the recent boom.[9] Nationwide, one bullish study estimated that the combined effects of a stronger oil and gas industry coupled with lower energy prices boosted economic activity by over $550 billion (about 3 percent of U.S. GDP) and created 4.3 million jobs.[10] At the regional level, another bullish study from Texas's Barnett shale region estimated that the shale boom was generating roughly $12 billion per year and over 100,000 permanent jobs.[11] A study of southern Texas's Eagle Ford shale estimated economic impacts of $87 billion and over 150,000 jobs.[12] Studies from other regions find similarly enormous economic impacts.

In many cases, industry advocates will cite these numbers as gospel,[13] ignoring the inherent limitations of any study that makes important assumptions about the way money cycles through the economy. As another note of caution, many of these studies rely on an analysis technique known as input-output modeling, which tends to overstate the economic and employment benefits of the industry because it ignores the possibility that increased activity in the oil and gas sector might negatively affect other parts of the economy.

For example, imagine you'd like to open an ice-cream shop in an oil and gas boomtown, but because construction crews are in high demand building housing for industry workers, hiring the labor to build your walk-in freezer costs twice as much as it would in a slower-growing economy. Similarly, hiring employees to scoop ice cream and press waffle cones is far more expensive than in other parts of the country. You consider passing these costs on to customers but suspect that no one will visit the shop if prices rise to $5 for a kiddie cup. As a result, the store never gets built. Research has indicated this hypothetical example has had real corollaries in booming parts of country in recent years, but like many other economic studies, the magnitude of the effects are uncertain.[14]

Adding to any skepticism over the accuracy of economic-impact analyses, the oil and gas industry or regional chambers of commerce have often sponsored these types of studies, and as one might expect, proindustry advocates sometimes inflate the numbers or state them as if they were pulled from God's very own accounting ledger.[15]

So while criticism of exaggerated claims can be justified, does that mean that the shale boom has done little for regional economies? Absolutely not. Some studies have used more careful modeling techniques to estimate the effect of the shale revolution. One particularly in-depth study estimated impacts across ten different plays and incorporated the effects of shale development on local wages and income, home values, and other factors. They also estimate some of the negative effects, such as increased traffic and crime (I explore these issues in chapter 11). Accounting for both benefits and costs, they estimate that on average, households in shale regions were better off by about $2,000 per year compared with those living outside of shale regions. They also found that these effects varied widely from place to place, with the costs outweighing the benefits in two regions (the Haynesville in Louisiana and the Woodford/Anadarko in Oklahoma) and the benefits outweighing the costs by as much as $9,000 per household in North Dakota's Bakken.[16]

Despite the difficulty of estimating economic impacts precisely, the central point remains: the recent boom has had a major effect on the economy of many regions of the United States. When we're talking about growth from 0.4 percent to 1.7 percent of GDP, we're talking big money, and those numbers have nothing to do with the methodological challenges of estimating overall impacts. Even though measuring these second- and third-order effects is difficult, the shale revolution has without a doubt rippled through the U.S. economy in a variety of important ways.

BONANZA

Drive into any oilfield town and you'll see them: the big white pickup trucks. You'll see them—usually Ford F-150s or some similar high-rider—at every stoplight. At every restaurant. Parked three deep in the hotel parking lot. Most of them are mud-splattered or caked in dust. Many are weighed down by heavy tow ropes or other equipment tucked into the bed liner. Some sport company names on their sides: Halliburton, Baker Hughes, or Airgas. Others are unlabeled: just another white truck hauling another worker to another job.

These trucks are a sure sign that you've entered oil and gas country. Arriving at hotels in rural northern Oklahoma, western North Dakota, or any other region frothy with drilling activity, big white trucks line the parking lots, arrayed like soldiers in formation. In places where drilling was slowing down, like southeastern Wyoming, eastern Utah, or northern Arkansas, the trucks would be sparse: one or two lonely soldiers stationed around back of the hotel.

The white pickups, more than the hundred-foot-tall drilling rigs or the massive natural gas–processing plants, signaled to me the presence of jobs. In a political era where "jobs" is the first, second, and third word out of every politician's mouth, the white trucks were proof that the oil and gas industry was, beyond any doubt, providing a lot of jobs. Take two examples. In Ellis County, Oklahoma, a rural county filled with rolling brush land abutting the Texas panhandle, a total of 271 new jobs were created across all economic sectors between 1998 and 2012. Of those new jobs, 258 came from oil and gas development. In 1998, there were around fifty oil and gas jobs in the county. In 2012, that number was greater than three hundred. In Williams County, North Dakota—the heart of the Bakken boom—jobs in the mining sector, almost all attributable to oil and gas development, increased from around three hundred in 1998 to over six thousand in 2013. The average oil and gas wage in Williams County was about $117,000 per year, compared to the U.S. average of around $50,000 across all industries.[17] There are plenty more examples where these come from.

The growth in oil and gas jobs has been especially beneficial for young men without college educations, who have suffered disproportionately through the recent recession. While many oil and gas jobs, such as working on a rig or driving an eighteen-wheeler, require substantial training, they don't necessarily require a college degree. And some, like the thousands of jobs that sprouted up working construction or staffing restaurants and hotels, required little training or education.

In many of the places I traveled to, these jobs were not abstractions. When traveling to the Bakken, Eagle Ford, or Permian, I would eat breakfast each morning in hotel lobbies that were often filled with young men in flame-retardant jumpsuits. Most days for lunch, the local diner or sandwich shop was crowded with the same. And at dinner time, I'd find myself waiting for a table or squeezing in to an empty barstool, where I'd sit alongside oilhands from out of town or longtime residents eager to tell me how much things had changed.

In other parts of the country, the local boom was less obvious, as I arrived well after its peak. In the Fayetteville (northern Arkansas), Haynesville (northeastern Louisiana), Green River (southwestern Wyoming), San Juan (northwestern New Mexico), and other regions, local residents and government officials told stories of the same type of boom sprouting in 2007, 2008, or earlier. By the time I visited in 2013, 2014, or 2015, many of the white trucks had driven elsewhere. Some were now plying the roads in new regions; others sat idle in storage yards or suburban parking lots. As I'll describe later in this chapter, the bust has in some regions hit just as hard as the boom.

In some of these places, especially those far away from the industry hubs of Texas, Oklahoma, Louisiana, and New Mexico, I could sense a trace of resentment among some residents over who had benefited most from the boom. In places like the Marcellus (Pennsylvania and northern West Virginia) and the Utica (eastern Ohio), much of the work had gone to newcomers from the southwest, identifiable to all by their license plates. These states have been trying to train more local workers to take advantage of the new oil and gas jobs, but progress has been slow, and the volatile cycles of the industry provide cautionary tales to those considering a career in the business.[18]

In a few places I visited, the boom was hardly noticeable at all. In the Barnett (northern Texas), home of the shale revolution, the endless sprawl of Fort Worth and Dallas subsumed the signs of the industry. Despite the dozens of rigs operating in the region when I visited in 2013, there was little in the way of visual cues to let you know that you'd entered oil country. I found a similar situation north of Denver, where dozens of rigs were tapping the Niobrara formation, but a dense population and diverse economy masked some of the economic effects of the boom. Where North Dakota felt like the Wild West, these places felt like everyday suburban America, albeit with the occasional drilling rig appearing on the horizon.

For other regions, the shale revolution was a life raft in stormy economic seas; that is, the timing of the boom allowed them to ride out the financial crisis and recession of 2008 and 2009 relatively unscathed. Although oil and gas drilling slowed sharply during the Great Recession, shale gas development was a boon to local communities during the deepest troughs of the crisis. In every major shale play, local residents and government officials told me they had barely noticed the broader economic downturn. While most of the country was struggling with excessive debt, collapsing housing prices, and a stock market in freefall, residents of shale regions such as the Bakken, Eagle Ford, Fayetteville, and Marcellus were buying new trucks, building new barns, and putting money aside for the future.

One study looked at how regional income and jobs were affected over the period of the recession and found that in shale regions, wages and other income increased by 54 percent compared with nonshale regions. They also estimated that during the Great Recession, the shale boom increased total U.S. employment by 640,000 jobs, translating to a decrease in the unemployment rate of 0.4 percent.[19]

And while jobs from the industry were a big part of the story, money flowed into these regions in a variety of other ways. The second key component—the one that most benefited longtime residents—was new income from leasing and royalties. Companies paid farmers and other landowners in shale regions anywhere from $500 to $20,000 per acre of land for the right to drill down to the shale below. Once drilling began, the residents would earn at least 12.5 percent of the revenues and

TABLE 9.1 **Economic indicators for Pennsylvania counties**

2007–2010 CHANGE	WAGES AND SALARIES	BUSINESS NET PROFITS	RENTS, ROYALTIES, COPYRIGHTS	TOTAL TAXABLE INCOME	TOTAL EMPLOYMENT
Most Marcellus wells	+2%	+14%	+461%	+6%	+1%
No Marcellus wells	−3%	−5%	+15%	−8%	−3%

Note: All percentage changes in real terms.
Source: With the exception of employment, these data are taken from K. Hardy and T. W. Kelsey, "Local Income Related to Marcellus Shale Activity in Pennsylvania," *Community Development* 46, no. 4 (2015). The employment data, which references the years 2007 through 2011, comes from T. W. Kelsey and K. Hardy, "Marcellus Shale and the Commonwealth of Pennsylvania," in *Economics of Unconventional Shale Gas Development*, ed. W. E. Hefley and Y. Wang (Switzerland: Springer International Publishing, 2015), 93–120.

sometimes as much as 20 or 25 percent. This income was a windfall for thousands. One study estimated that between 2005 and 2013, the six biggest shale plays generated roughly $39 billion in royalties for private landowners.[20]

A series of studies from researchers at Penn State University documented the rapid growth in revenue from royalties and other sources.[21] In the twelve Pennsylvania counties with more than ninety Marcellus wells, the researchers found that almost every economic indicator dramatically outperformed the state average. And when compared with the thirty-one counties with no new wells, the gap in economic performance grew further. Most strikingly, as table 9.1 shows, income from rents, royalties, and copyrights exploded by more than 450 percent in Marcellus counties, indicating large bonus and royalty payments for local landowners. But the economic benefits did not just accrue to property owners. Counties with the most Marcellus wells also experienced far stronger economic performance in wages and salaries, business net profits, total income, and total employment.

THE DOWNSIDE OF THE BOOM

For the most part, the economies of these regions benefited enormously from the boom (I'll discuss the effects of the bust later in this chapter). However, a word of caution is warranted. A growing economy, new job opportunities, and increased income are clearly good things for a community, and most of the people I met across the country were beneficiaries of this growth. However, there are often members of the community who for a variety of reasons do not benefit and who may in fact suffer because of increased economic activity, whether generated by oil and gas development or any other fast-growing industry.

For example, fracking booms have led to rapid population growth in some rural areas, such as the Bakken, Eagle Ford, Marcellus, and others. This growth can cause housing rents to skyrocket, sometimes growing by 400 percent over the course of a year or two.[22] Consider this: residents of San Francisco have complained for years about being priced out of their homes by the nouveau riche of the tech industry.[23] But in 2014 it was Williston, North Dakota, that ranked as the most expensive place to find rental housing in the United States.[24]

For longtime property owners in places like Williston, growth in rents is fantastic news. If you owned a storefront that had sat vacant or underutilized for years, the prospect of renting the space for thousands of dollars per month would be manna from heaven. Similarly, many homeowners in the region rented out spare rooms to oil and gas workers willing to pay top dollar for a decent place to sleep.

But not everyone owns property, and not everyone can take advantage of the boom. Instead, they may suffer as the cost of living rises. Imagine a retiree living on a fixed income in a small apartment. Skyrocketing rents mean she'll either need to find a crummier place to live or move out of the area entirely. And in many cases, those who can't afford increased prices are those who have the hardest time finding new sources of income or relocating to another part of the country. When McDonald's pays fifteen dollars per hour for cooks and cashiers, those wages are wonderful news for a young worker without a college degree. But for a disabled person relying on a monthly Social Security check, higher prices for a cheeseburger aren't easy to swallow.

In North Dakota and Pennsylvania, I spoke with several longtime residents who were retired and struggling to keep up with the increased cost of living. Both states implemented plans to subsidize more low-income housing in booming areas, using portions of the revenue derived from the oil and gas industry to build new apartment buildings. But the folks I met had not benefited from those housing programs. And while they were happy to see their community thriving, it was clear that the shale boom was not a boon to all.

CHEAP GAS! CHEAP OIL?

As significant as the fracking boom has been for the hundreds of communities I've visited over the past few years, perhaps an even larger economic effect has been the national and international consequences of the shale revolution: cheap natural gas and, in a more complicated sense, cheap oil. Let's start with natural gas.

For most of the 1990s and 2000s, policy makers, economists, and others were worried about the coming shortage of natural gas in the United States (as discussed in chapter 8). Production was essentially flat, and consumption was growing rapidly. Natural gas imports through pipelines from Canada were becoming a large part of the domestic supply, and most everyone expected that we were going to need to import lots more of the stuff. Companies invested billions of dollars in building enormous LNG-import terminals capable of accepting gas transported by specialized ships from places such as Russia and Qatar.[25] Federal-government projections showed natural gas prices staying high for the foreseeable future, which meant higher heating, cooking, and electricity bills. Industries that rely heavily on natural gas, such as plastics, fertilizers, and other manufacturers, looked abroad to regions with lower energy prices in which to build their new factories.

But as shale gas production boomed, natural gas prices started to fall. In 2008, Americans spent about $230 billion on natural gas. In 2010, after shale gas (coupled with the Great Recession) drove prices down, Americans consumed more natural gas than in 2008 but spent about $70 billion less.[26] That savings of $70 billion equates to $226 per person. Winter heating costs, which include natural gas, electricity, and heating oil and can be a substantial burden for those living in northern climes, fell from about $900 per year during the 2008–2009 winter to about $600 per year in the 2011–2012 winter. What's more, these low energy prices disproportionately benefit low-income households, who spend a larger share of their income on energy than the wealthy.

The shale boom has also helped drive oil prices from their hundred-dollar-levels to lows of less than $35/barrel in 2016. The effects of fracking on oil prices are complicated, because unlike natural gas, oil is widely traded on a global market. The fracking-driven oil boom in the United States is not the only reason for the oil-price crash that began in late 2014, but it was an important piece of the puzzle.[27] Other factors, in particular slower economic growth in China and the decision by OPEC (led by Saudi Arabia) to pump at full tilt, played major roles. However, the price probably would not have fallen as far and as fast as it did without the roughly four million barrels of new oil the United States pumped out of the ground each day in 2014 and 2015. In 2017, as OPEC cut production, surging shale from the United States continued to dampen prices.

For U.S. consumers, the oil crash has pushed down prices in all sorts of ways. The first thing drivers notice is a sharp drop in gasoline prices. Because of lower oil prices, total domestic spending on gasoline and diesel (the two leading transportation fuels) shrank from $707 billion in 2014 to $519 billion in 2015, a decline of $188 billion, or $1,600 per household. This despite the fact that drivers guzzled about 2.7 billion gallons more of the two fuels combined. If households spend some of that saved $188 billion, the economy gets a jolt, and the benefits are multiplied.

In addition, low natural gas and crude oil prices reduce the costs of all sorts of other goods. The cost of food goes down because farmers pay less for natural gas–derived fertilizers and less on fueling their tractors and combines; the cost of consumer goods goes down because boats, trains, and trucks spend less on fuel; and the costs of plastics goes down because oil and gas are the main feedstock for plastic producers.

But the rapid growth of the domestic oil and gas industry may complicate the simple narrative that lower prices benefit the economy as a whole. For decades, most researchers have believed that, because the United States consumes much more oil than it produces, the economic benefits of lower prices for consumers will tend to outweigh the shrinking incomes for producers.

However, recent research has provided evidence that from 2014 to 2016, the consumer benefits were essentially cancelled out by a decline in investment in the oil and gas sector, resulting in a net impact to GDP of roughly zero.[28] With a larger domestic oil and gas industry, the downsides of a sharp drop in oil prices multiply. Intuitively, lower prices for oil and natural gas are particularly painful for the economies that rely on pulling them out of the ground in places such as Alaska, Texas, or North Dakota.

THE HAPPY TAXMAN

The shale revolution hasn't just affected the private sector. It's also done an enormous amount to raise money for governments in oil- and gas-producing regions. Increased GDP, employment, royalty payments, and all the other economic effects of the boom translate into increased tax revenue, which helps pay for schools, roads, and other government programs (of course, the benefits can disappear quickly in a bust, as I'll discuss). While state and local governments feel these effects most acutely, the federal budget has also benefited.

Increased oil and gas production has boosted federal revenue primarily through the broad economic effects of the boom. More jobs with high wages lead to more income taxes, and higher income for businesses generates additional corporate taxes. In a 2014 study, the Congressional Budget Office estimated that shale will make federal revenue higher than it would have been by 0.8 percent in 2020 and 1.0 percent in 2040.[29] This might not sound like a lot, but 0.8 percent of federal revenue in 2020 would equal roughly $35 billion. That's almost enough to fund the Department of Energy ($30 billion in 2015) and the EPA ($8 billion in 2015) combined (of course, these effects will be higher or lower depending on the level of industry activity across the country).[30] Whether policy makers in 2020 or 2040 will be funding government agencies to this level is an open question, but the notion that a stronger oil and gas industry will provide additional revenue is clear.

Another helpful revenue source for the federal government comes from the money paid by companies to explore for and produce oil and gas on federal lands. In 2014, onshore oil and gas production on federally owned land generated roughly $3 billion in royalties.[31] While fracking was not responsible for most of this revenue, it will become an increasingly important contributor to federal coffers in years to come, particularly if access to drilling is expanded under the Trump administration. Of course, there is much debate over how widely federal (and state) lands should be opened up to oil and gas production, with proponents highlighting the potential economic benefits and opponents focusing on the risks of environmental degradation.

Whether or not the federal government expands access, some research has suggested that it could be getting more bang for its buck (such as through higher royalties from its offshore and onshore oil and gas fields).[32] In addition, research suggests that the $3 to $4 billion in federal subsidies for oil and gas producers embedded in the U.S. tax code benefit smaller oil and gas producers but do little to improve domestic energy security or mitigate climate change.[33] If these policies were changed, the financial benefits of fracking for the feds would grow further. Nonetheless, the central point is hard to dispute: shale development has boosted, and will boost, federal revenues for years to come.

State and local governments have also seen swollen budgets thanks to fracking. In Texas, the largest shale producer, revenue from the state's oil- and gas-production taxes grew from about $1.5 billion in 2003 to $4.6 billion in 2014 thanks to shale development (2015 revenues were roughly $4 billion).[34] In North Dakota, the state with the highest rate of production growth, revenue from the state's two major taxes on oil producers grew from around $75 million in 2004 to more than $3 *billion* in 2014. This enormous surge in cash meant that oil money accounted for more than half of the state's entire tax revenue in 2014.[35] Revenue in states such as Wyoming, Colorado, Louisiana, New Mexico, and others tell similar, if less dramatic, stories. Some states, notably Pennsylvania and Ohio, tax oil and gas production relatively lightly and have seen less in the way of tax-revenue growth.

Figure 9.1 shows the amount of revenue raised by state and local governments from the oil and gas industry in fiscal year (FY) 2013. The largest single revenue source is severance taxes, so-called because they tax minerals that are "severed" from the ground. In major producing states, severance taxes generate hundreds of millions or billions of dollars each year. Some states, such as Arkansas, California, Ohio, and Utah, collect little in the way of severance taxes because they have low severance-tax rates.

Like the federal government, states also lease land for oil and gas production, generating billions more each year. Because the state government owns the land, it is entitled to bonuses and royalty payments just like private landowners. Like severance taxes, state leases can generate hundreds of millions or billions of dollars each year, though they also

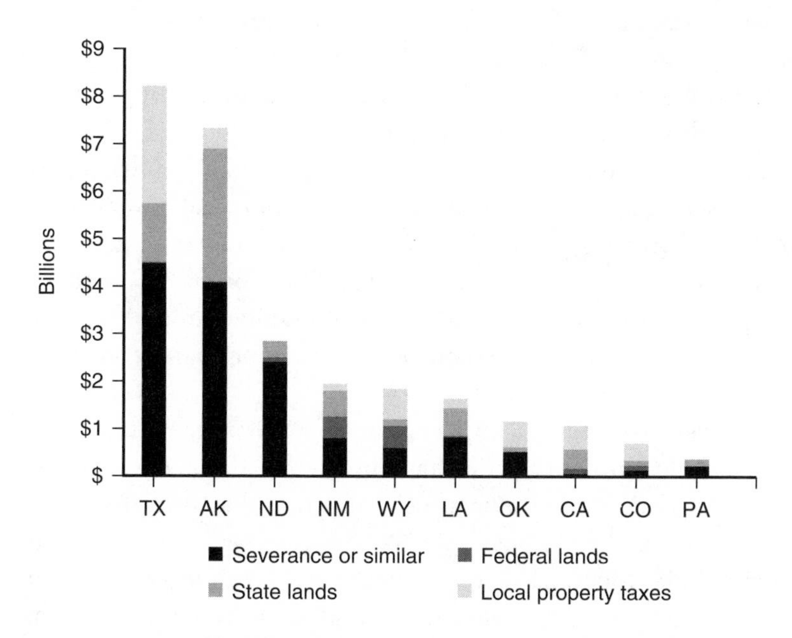

FIGURE 9.1 **State- and local-government oil and gas revenues, FY 2013**

Source: D. Raimi and R. G. Newell, "U.S. State and Local Oil and Gas Revenues," Resources for the Future Discussion Paper (2016), 16–50.

vary widely. For example, most oil production in Alaska occurs on state-owned land, generating billions of dollars in revenue. The state, in turn, invests a portion of those revenues into an investment fund that doles out annual checks to every Alaskan.[36]

In many states, especially those in the West, revenue from state leases is dedicated to permanent education funds, which help fund schools for future decades. For example, Texas's university system owns millions of acres of land and invests hundreds of millions of dollars in annual oil and gas revenues from those lands into a permanent fund that supports universities across the state. As a result, the University of Texas's endowment ranks alongside those of Harvard and Yale as one of the largest in the United States.[37] Similarly, western states such as Colorado, Montana, New Mexico, North Dakota, Utah, and Wyoming invest virtually all of their state lease revenues into education trust funds. In total, U.S. states invested more than $2.5 billion into those funds in FY 2013.[38]

Local governments have also largely benefited from fracking, though they face a number of challenges as well. In most parts of the country, local governments have seen taxes and other revenue sources rise dramatically when fracking comes to town.[39] One of the leading sources, for schools in particular, comes from property taxes. Just like homes, factories, and other property, oil and gas wells, processing plants, and other infrastructure are subject to property taxes from local governments. In FY 2013, local governments, including schools, counties, and cities, collected roughly $5.7 billion in property taxes from oil and gas producers.[40] This revenue has enabled hundreds of communities around the country to build new parks or pools, improve their roads, or upgrade local schools because of the fracking boom.

However, growing oil and gas production also presents challenges for local governments. In particular, a boom can drive rapid growth in population and vehicle traffic, which can strain local infrastructure, law enforcement, and other public services. In some big shale states, namely, Pennsylvania, Montana, and North Dakota, local governments aren't allowed to collect property taxes on oil and gas wells or other equipment. Instead, the state government collects a tax or fee on oil and gas production, then sends some of that money back to the local level. In some cases, especially North Dakota and parts of Montana, these revenues haven't always been enough to keep up with demand for things like road repairs and other infrastructure. As a result, some local governments have needed to take out hundreds of millions of dollars in loans to keep up with a rapidly growing population.[41]

The second issue is related: local governments can be overwhelmed by a fracking boom. It hasn't happened in many places, but when it does, some of the people who have lived in communities for a long time see their public services suffer, their cost of living rise, and their sense of community untangle. For example, the boom that started in Rifle in the 2000s created major economic growth, accompanied by major challenges for the local government. The city suddenly saw thousands of people moving into the area to work in the gas fields and scrambled to build new water lines and sewer facilities, the types of projects that don't generate many headlines but that cost tens of millions of dollars. Rifle invested heavily in these upgrades, expecting the new residents to pay for them with their sewer and water bills over the decades to come.

But at the end of 2008, the music suddenly stopped. The national economy collapsed, along with natural gas prices and the local housing market. The rigs went back to their storage yards in Texas and Oklahoma or moved to more profitable shale plays in Pennsylvania and North Dakota. Most of the workers went home, too. The thousands of new residents the city was banking on to pay for those new water-treatment plants were gone, and construction was already underway. What now?

In the end, Rifle had to double its water rates and increase its sewer rates by about 50 percent. The city raised sales taxes, too, in an effort to stabilize local-government finances. The people who had lived in Rifle for decades, along with the relatively few who remained from the boom, had to pay for the bust.[42]

BOOM AND BUST

The story of public services in Rifle illustrates a broader issue. Through 2014, most of the economic news related to oil and gas production in the United States was rosy, bordering on giddy: thousands of new high-paying jobs, regional economies in places like Texas and North Dakota going gangbusters, oil and gas prices falling, and everyone's happy. But the good times don't last. In the oil and gas business, they rarely do. As anyone who's been around the industry for more than a few years knows, the boom is always followed by a bust. And if you put all your eggs in the oil and gas basket, you'll be in for hard times when that day inevitably comes.

Because prices for oil and natural gas don't move in lockstep in the United States (though they do in some other parts of the world), booms and busts for oil and natural gas can happen at different times. From 2005 through 2008, there were dozens of rigs roaming the mountains and mesas surrounding Rifle, drilling into the tight sands to access high-priced natural gas. But when the price of gas fell from about $12 per million British thermal units (MMBtu) in the middle of 2008 to less than $4/MMBtu a year later, only a few rigs remained there, targeting just the richest spots that could still generate a profit. Natural gas booms in other parts of Colorado, Wyoming, and Utah stalled or began to backtrack, taking with them thousands of workers and the economic activity they generated. When the price recovered somewhat a few years later, companies eyeing natural gas turned primarily to shale plays in places such as Pennsylvania, Louisiana, and Arkansas, while tight sands development in Colorado, Wyoming, and Utah remained anemic.

At the same time, many other drillers turned away from natural gas entirely, instead shifting their focus to plays with more oil (often referred to as "liquids-rich" plays).[43] Oil prices hovering around $100/barrel in 2011 and 2012 meant lots of profitable drilling opportunities, which meant lots of rigs, rig workers, truck drivers, pipeliners, and more. But oil prices can be just as volatile as natural gas prices, as these companies would soon be reminded.

Beginning around 2008, the biggest and longest-lasting booms took place in liquids-rich plays, particularly the Bakken, Eagle Ford, and Permian basin. Together, these three regions were producing more than 4 million barrels of oil per day by 2014, accounting for almost all of the growth in domestic production. These booms transformed the cities and towns of North Dakota's Bakken and southern Texas's Eagle Ford, while the Permian basin, a region well acquainted with booms and busts, saw a new cycle of roaring economic growth.

But in late 2014, as U.S. oil production continued surpassing all expectations, the market began to shift. Along with the unprecedented growth in U.S. supplies, growing oil production from other nations, particularly Saudi Arabia, coupled with slack global demand to burst the bubble. Prices fell by more than half over the span of a few months in late 2014 and early 2015, leading companies to slow drilling, lay off workers, and, for many, file for bankruptcy.

According to one law firm with a large energy division, over one hundred oil and gas companies entered into bankruptcy between early 2015 and late 2016.[44] In just one year, the share of GDP from oil and gas extraction fell from 1.7 percent in 2014 to 1.0 percent in 2015. Employment in the sector dropped by more than 20,000, falling from more than 200,000 in 2014 to 177,000 by the summer of 2016.[45]

And while the bust in oil and gas prices has been a windfall for American consumers, it can be devastating to the industry. As noted earlier in this chapter, some researchers have estimated that the reduction in investment by oil companies more or less canceled out the economic gains from lower prices.[46] That reduced investment, in turn, slows down activity in the oilfield, meaning fewer jobs across oil-producing regions. What's more, the price crash led companies to look for every opportunity to cut costs. And like advances in other industries, improved automation technologies in the oilfield have reduced the demand for workers to operate machinery.[47]

Looking at the issue at a more personal level, two recent studies have found that high-paying jobs in the oil and gas industry led to an increase in students dropping out of high school.[48] While dropping out of school doesn't sound like a great strategy for long-term prosperity, it probably made a lot of economic sense at the time. Instead of sitting in class, students could drive a truck or work on a rig to earn $80,000 or $150,000 per year. This type of job could be a great option for students that are interested in something other than college or office work. Manufacturing employment has decreased steadily in the United States over the past few decades, and oil and gas jobs could help some of the folks displaced by the decline of those industries.[49]

There's just one problem: What do you do when the bust comes? Once that high school dropout loses his job on a rig, he may be twenty-one years old. Is he going to head back to high school? Probably not. Instead, he may have bought a truck, moved into a house, and taken out a mortgage premised on an annual salary of $100,000 or more. Now, he has no prospects for work. He could sit at home and wait for oil and gas prices to come back up, but who knows when that will happen?

Oil and gas prices are volatile, and when they drop, the industry responds quickly. The prospects for an oil- and gas-rig worker to secure a long-term, stable, middle-class life can come and go as quickly as the prices flashing across the digital displays of the New York Mercantile Exchange. Figure 9.2 shows the number of rigs drilling for oil in the United States growing steadily from mid-2009 through the end of 2014, as oil prices rose from their depths during the Great Recession. As prices hovered in the neighborhood of $100 per barrel from 2011 through 2014, rig counts—and jobs in the oilfield—kept climbing.

But when prices crashed at the end of 2014, so did the number of rigs out in the fields. If you were a worker on one of the more than eight hundred rigs that suddenly went idle, you may have been working in the oilfield for one, two, or three years. If you were one of those who dropped

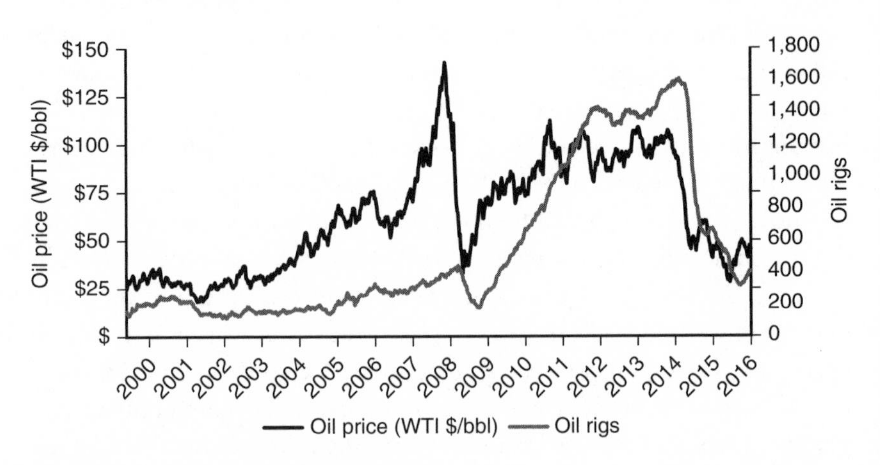

FIGURE 9.2 **Oil prices and rig counts**

Sources: Baker Hughes, "Rig Count Overview & Summary County" (2016), http://phx.corporate-ir.net/phoenix.zhtml?c=79687&p=irol-rigcountsoverview; U.S. EIA, "Petroleum and Other Liquids Prices" (2016), https://www.eia.gov/petroleum/data.php#prices.

out of high school or college and never experienced a bust before, you may have had no idea it was coming.

I met dozens of these men (they are mostly men) at restaurants and bars in places such as Watford City, North Dakota; Odessa, Texas; Greeley, Colorado; and McAlester, Oklahoma. They worked on rigs, drove trucks, built pipelines, ran frack jobs, and more. You could usually identify them by their well-worn Carhart overalls or red Halliburton jumpsuits. In 2013 and 2014, many didn't have any qualms about ordering the biggest steak on the menu or buying a round of shots for friends (and the occasional policy researcher). But in 2015, as oil prices crashed and thousands of rig workers were laid off, there were suddenly fewer big spenders out on weeknights. I wondered how many of the workers I'd met in previous years had saved some of their windfall.

More broadly, economies built on oil, gas, or other extractive industries can be subject to something called the *resource curse*. A key element of the resource curse is that if regions or nations become overly reliant on a finite natural resource like oil and gas, the economy will fail to develop other sectors because so much attention and money is channeled toward extracting the resource (recall the ice-cream shop example from earlier in this chapter).[50] In these regions, where extraction is the only game in town, a bust in commodity prices can drive a stake through the heart of the local economy.

However, it's far from clear whether the resource curse applies in the United States (most research on the topic examines governments and economies in nations such as Nigeria, Russia, Saudi Arabia, or Venezuela). In some regions, like the Permian basin, most of the economy has been underpinned by oil and gas extraction for decades. In the Bakken or the Eagle Ford, where oil and gas had not played as big of a role, local officials have worked to stave off the resource curse by trying to diversify their economies, even when oil prices and production were high. In a worst-case scenario, years or decades of low oil prices can impoverish a community that has failed to diversify. As economic opportunity dwindles, the community may begin to empty out and perhaps even become a ghost town like those found in the Old West, casualties of previous mining booms.

As noted toward the beginning of this chapter, there's mixed evidence about whether oil- and gas-driven economies in the United States are subject to the resource curse. That is, if the boom had never happened, would they be better off or worse off in the long run?

One thing is certain: busts are painful. In Rifle, the restaurants were empty, and only a few white pickups occupied the parking lot at the Holiday Inn. In Midland, government officials described how during the 1990s, when oil prices stayed low pretty much the entire decade, the city dried up, and gleaming office buildings turned to dusty eyesores. In Williston, the cost of a night's sleep at local hotels dropped from around $300 to $100 per night—good news for a researcher on a tight budget but a bad sign for the city's economy.

SUMMING UP

The shale revolution has generated large numbers of jobs and spurred substantial economic growth, but those benefits have been volatile. In many parts of the United States where oil and gas drilling has grown rapidly, the economic boom has been impossible to miss. However, the volatile nature of oil and gas prices means that these communities are subject to "booms and busts" and that the economic benefits for an economy reliant on the industry can disappear rapidly.

At the national and international scale, fracking has had a major effect on all energy consumers by contributing to lower-cost oil and natural gas. However, a larger domestic oil and gas industry means that low prices take a larger toll on producers, reducing—and potentially negating—the overall economic benefits of the boom.

Broadly speaking, oil and gas production is not an economic panacea, and it is more subject to booms and busts than almost any other industry. For regions that have experienced rapid growth thanks to fracking, it's important to appreciate the economic benefits the boom has brought, but it's just as important to prepare for the day when those benefits will disappear.

10

WILL FRACKING SPREAD AROUND THE WORLD?

The advances in technology that enabled the shale revolution were developed primarily in the United States, but the source rocks that produce this oil and gas can be found all over the world. In fact, the United States isn't even the largest holder of shale gas resources. An assessment released in 2013 and updated in 2015 estimated that other regions hold as much shale gas potential as the United States, if not more (see figure 10.1). The assessment also showed that tight-oil resources are abundant around the globe. And these estimates do not include some of the world's biggest producers, such as Saudi Arabia, Iran, or Iraq, which likely hold vast potential as well.

But despite the widespread presence of oil and gas trapped in shale formations around the world, and despite the fact that the United States has been producing shale gas in large quantities for over ten years, no other countries have made major headway into developing their own resources.

That the shale revolution occurred in the United States, not elsewhere, was far from dumb luck. A mix of federal-government policies, oil- and gas-industry entrepreneurship, and the United States' unique structure of private land ownership worked together to encourage these new technologies, then helped spread them quickly across the country.

While the oil and gas industry has at times seen the U.S. federal government as an overbearing scold, impeding production with a litany of rules and restrictions, it was the federal government that helped lay the groundwork for shale production by supporting early research and

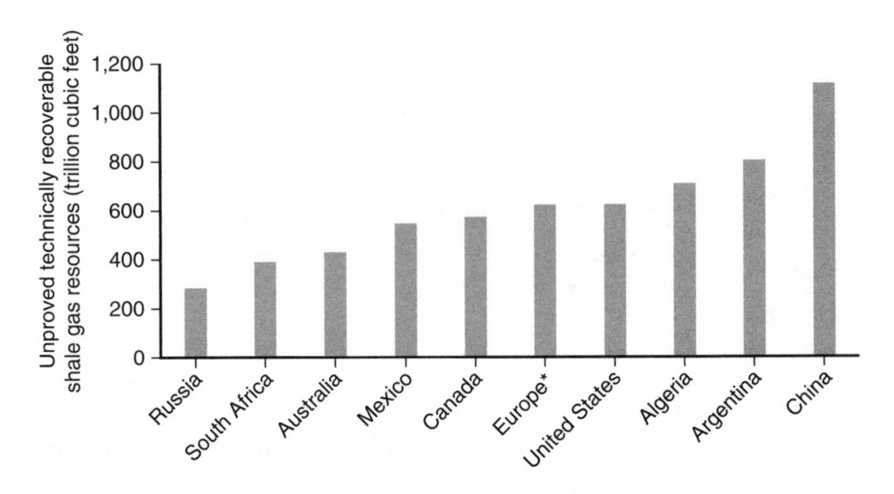

FIGURE 10.1 **Global shale gas resources (select regions)**

*Includes Bulgaria, Lithuania/Kaliningrad, Poland, Romania, Turkey, Ukraine, Denmark, France, Germany, Netherlands, Norway, Spain, Sweden, United Kingdom.

Source: U.S. EIA, "World Shale Gas Resources" (2015), https://www.eia.gov/analysis/studies/worldshalegas/.

development. As I described earlier in the book, this support ranged from the mundane (such as offering tax credits for the production of "unconventional" natural gas) to the eye opening (such as the support for nuclear fracking tests in Colorado and New Mexico). The foundational knowledge developed through this research helped spur the industry in its early days.[1]

Another key factor is the entrepreneurial soul of the U.S. oil and gas industry. Risk-taking by nature, for most of its history, more than 50 percent of new wells came up "dry"—that is, more than half failed to produce enough oil or gas to make a profit and had to be sealed and abandoned. Today's Silicon Valley culture may pride itself as a believer in failing fast, then learning from one's mistakes, but the oil and gas industry has been doing just that for over one hundred years.

Following in that tradition, George Mitchell and his company drilled for gas in the Barnett shale long after most businesses would have folded their tents. But Mitchell was hardheaded—he was convinced his team could crack the code. And with each new well, his petroleum engineers learned more, eventually finding a way to make shale gas profitable.

And once they cracked the code, the secrets of shale development spread rapidly throughout the industry. Mitchell's company was bought by Devon, a larger natural gas producer based in Oklahoma City. Soon other companies, such as Chesapeake, Southwestern, EOG, and Continental, began drilling into shale themselves. These companies, some of the earliest movers in the shale revolution, benefited from the close-knit nature of the oil and gas industry and were willing to lay their big money on the line to make shale development a large-scale endeavor.

But these two elements—a firm foundation of knowledge supported by government-backed R&D, paired with an entrepreneurial "wildcat" spirit—wouldn't have been enough without the third piece. And while it may sound like a technicality or something out of a first-year class in law school, this third piece is what distinguishes the United States from every other nation on earth, and it has played an enormous role in the rapid spread of fracking: private mineral ownership.

The concept of private mineral ownership—that private citizens own not just the surface of their property but also all the stuff underneath—is essentially unique to the United States. In other nations, the government owns the subsurface, including any valuable minerals that may be found there. Why does that matter? Say you live in Anytown, U.S.A., on a fifty-acre plot of land that just so happens to sit atop a large shale deposit. A company approaches you about drilling on your land and, in exchange, offers you $1,000 per acre for the right to drill, plus 15 percent of the value of oil and gas produced from underneath your property. You may be concerned about the environmental hazards and disruption to your daily life, but you'd probably think hard about the prospect of all that cash.

Now imagine you live on that same plot of land in a country where the government owns the mineral rights. In this scenario, you are still subject to the environmental risks and day-to-day hassles that come with an oil well in your backyard, but because you don't own the minerals beneath your land, you have no opportunity to profit from their extraction. Instead, the oil company will pay the *government* that $50,000 for the right to drill, plus 15 percent of the revenues that come in.

While there are exceptions, private landowners in the United States have in most cases enthusiastically welcomed this opportunity, helping the industry expand rapidly. In the heady days prior to the Great Recession, as Chesapeake Energy and other companies sought to gather as much land as they could, landowners that lived atop shale plays like the Haynesville in Louisiana were seeing bonuses of more than $20,000 per acre with royalties as high as 20 percent.[2] In other nations, where landowners have far less to gain and just as much to lose, local residents are understandably more prone to skepticism about the industry moving into town.

And while this key difference has helped slow the spread of fracking globally, shale development is moving forward—in some cases at a gallop, in others at something less than a trot—in a number of other nations.

The political, social, and geological specifics of each country have led to a variety of different policy approaches toward shale development. While some nations pursue it at full speed, others have been reluctant to embrace fracking, primarily driven by concerns over its environmental risks. Some countries in Europe, including France, Germany, and Bulgaria, have imposed moratoriums or bans on fracking, though they continue to permit "conventional" oil and gas development. Other European nations, such as Poland and Spain, have been more open to tests and exploratory drilling of shale wells, though neither country has seen large-scale production.

Several African nations have moved forward with experiments and small-scale shale development, but fracking has not spread rapidly there, either. For example, Algeria is already a major natural gas exporter to Europe and other nations, but environmental concerns, limited foreign investment, and other factors have prevented it from developing its large shale gas resources.[3] In South Africa, where coal is the dominant fuel, shale gas resources are also abundant, and the government has encouraged development after drafting rules in 2011 and 2012. While major oil companies such as Shell and Chevron have looked to drill in the country, there is substantial local concern about environmental impacts in the aridly pristine Karoo region, where most of the drilling would occur.[4]

While these countries offer their own lessons into shale development, three other countries I've visited in recent years—China, Argentina, and the United Kingdom—offer particularly relevant insights. In these three nations, governments, companies, and citizens have moved forward on fracking with a mix of hope, opportunism, and skepticism. In each case, there is no doubt that vast quantities of natural gas (and, in some cases, oil) are trapped in shale rocks below the surface. A 2016 projection from the U.S. Energy Information Administration estimated that by 2040, roughly 30 percent of the world's natural gas would come from shale.[5] But projections are always uncertain, and while it seems likely that fracking will move forward in each country, the speed and scale of that development remains very much in doubt.

SICHUAN, CHINA

Sichuan province, located in southwestern China, is perhaps best known in the United States as a home of amazingly eclectic and often spicy cuisine (think of the tongue-numbing Sichuan peppercorn). Abutting the edge of the Tibetan plateau, it is home to more than 87 million people, more than twice the population of California. It also contains substantial shale resources.

With China's astounding economic growth through the 1990s and 2000s came a commensurate increase in demand for energy, met mainly with coal. From 1990 to 2015, China's coal use more than tripled, and by 2015, it burned almost five times as much coal as India, the second largest consumer.[6] This enormous rise in coal consumption helped industrialize the Chinese economy, but it has also contributed to hazardous levels of local air pollution and a rapid rise in greenhouse-gas emissions. In the mid-2000s, China rocketed past the United States as the world's largest emitter of carbon dioxide, primarily because of its increased coal consumption, and today it emits more than any other nation by a wide margin.

To reduce local air pollution and meet internationally agreed-upon greenhouse gas–reduction targets, the Chinese government has taken action to reduce its reliance on coal-fired power. Along with supporting increased deployment of nuclear, hydro, wind, and solar, the government has provided substantial incentives to increase domestic gas production. And China's potential for natural gas is enormous. Along with substantial "conventional" natural gas fields, China has some of the world's most promising coalbed methane resources.[7]

But it's the shale in Sichuan that has sparked perhaps the greatest interest from the Chinese government and from international oil companies. To spur investment, the government in 2012 promised to pay an extra $1.80 per thousand cubic feet for natural gas produced from shale formations—not a shabby subsidy, given that gas in the United States was in some 2012 months selling for around the same amount.

As companies have learned more about the resources in the Sichuan basin, the costs of drilling new shale wells have decreased substantially,[8] and production has grown—tripling, for instance, from 2014 to 2015. Still, China has a long way to go.[9] In 2015, its shale wells produced roughly 500 million cubic feet of shale gas per day, about 4 percent of what Pennsylvania produced that same year.[10]

There are several factors behind this slow growth. First, China's state-run oil and gas companies generally do not have the same entrepreneurial and risk-taking culture found in the U.S. oil and gas industry, and there are far fewer companies with the expertise needed to develop shale wells. In addition, the pipeline system that would transport gas from Sichuan to major consuming centers is limited, and not all companies have access to those pipelines, making it harder for new companies to break into the market. Other major constraints include a relatively limited supply of fresh water and a complex land-ownership system.[11]

Despite these challenges, rapidly growing demand for natural gas will continue to encourage the Chinese government (and perhaps private companies) to develop shale plays like the one in Sichuan. From 1990 to 2010, China's natural gas consumption grew by more than 500 percent. Over the next twenty-five years, China's demand for gas is projected to more than triple under the International Energy Agency's "New Policies Scenario" (essentially, the agency's best guess about the future). Even in a "450" Scenario, where the world limits its greenhouse-gas emissions

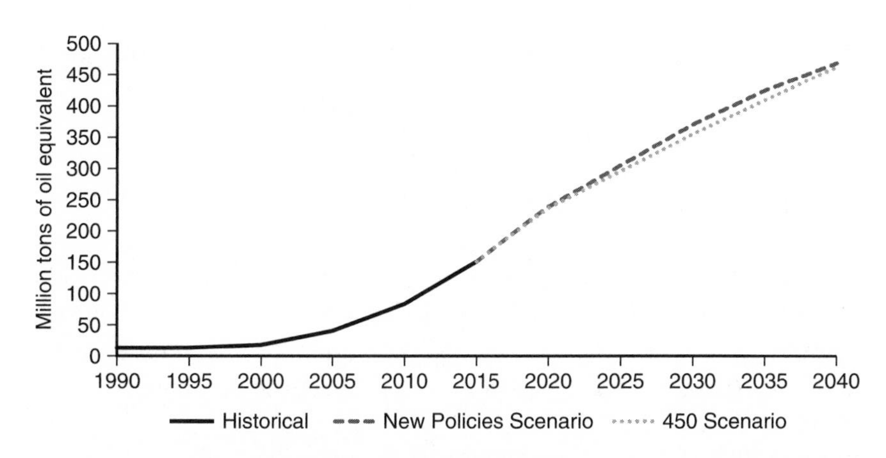

FIGURE 10.2 **Natural gas consumption in China**

Source: International Energy Agency, World Energy Outlook 2016, http://www.iea.org/newsroom/news/2016/november/world-energy-outlook-2016.html. See tables in appendix A for projections on Chinese natural gas consumption.

dramatically,[12] Chinese natural gas consumption grows at roughly the same rate over the next several decades, displacing even more coal to reduce emissions (see figure 10.2).

In short, shale gas development has moved more slowly than some had expected in China over the past several years. But where there's demand, oil and gas companies often find a way to produce the needed supply. However, two key questions remain. First, will China be able to attract private investment, given the large role the government plays in the oil and gas sector? And if not, will China's large state-controlled oil and gas companies be nimble and innovative enough to tap the country's vast resources?

Fu Chengyu, a former chairman of Sinopec, one of China's state-owned oil and gas companies (and according to some reports the world's fourth-largest company)[13] is optimistic about the prospects for shale gas in China. In an interview discussing prospects for the country's energy future, he states that natural gas (along with renewables) is poised to grow quickly and displace coal and that "In the past we've had pipes importing gas, and we've had LNG gas, and also domestic production gas.

For domestic production, we see that shale gas will become the major driver for gas supply in China."[14]

THE UNITED KINGDOM

On a 2016 visit to the United Kingdom, shortly before voters opted for a "Brexit" from the European Union, I found myself on a train journey from Swansea, Wales, to Oxford, England. Along the way, acres of emerald-green hedgerows stretched out in neat patterns, interrupted occasionally by the odd collection of sheep, cows, or horses.

Standing out from these bucolic scenes as we neared big cities like Bristol were the occasional and unmistakable shadows of twenty-story cooling towers, the massive concrete cylinders where water used in large power plants is cooled. While these cylinders are typically associated with nuclear plants (picture the two concrete towers at Homer Simpson's workplace in Springfield), they are also a fixture of many coal-fired power plants.

As early as the 1200s, Great Britain's coal—originally called *sea-coal*, most likely because it was transported around the country via waterways—provided heat for homes in London, displacing the burning of wood from England's rapidly shrinking forests and spewing noxious fumes into the city air. By the late 1700s, the coal industry had grown enormously, fueling the steam engine and the subsequent Industrial Revolution. By the 1830s, Britain produced about 80 percent of the world's coal, and by the 1850s, coal helped the United Kingdom produce more iron than the rest of the world combined.[15]

While much of this coal was mined in and around the northern city of Newcastle, Welsh mines also contributed a substantial portion (I say with pride: my great-grandfather worked for a Welsh mining company). This Welsh coal was then shipped east on the very railroad tracks I was traveling along, helping power England's great industrial cities like Liverpool and Manchester, or farther south to the cacophony of Dickensian London.

By the twentieth century, coal had new competition: oil. In the run-up to World War I, Winston Churchill—at that time head of the British navy—transitioned the royal fleet from coal to oil, speeding along the century's embrace of petroleum. But unlike coal, oil was not plentiful in England. It relied on the United States and, later, Middle Eastern nations such as Iran and Iraq to quench its thirst for oil-based fuels.[16]

Then in the 1960s, the United Kingdom made a major find of its own. First came the discovery of large quantities of natural gas beneath the North Sea, a cold and harsh expanse off of England and Scotland's east coast, bounded to the east by Norway and to the south by Germany and the Netherlands. Several years later, large oil deposits were discovered beneath the North Sea. Over the next thirty years, dozens of new fields were found and tapped, transforming the United Kingdom (and, to an even greater extent, Norway) into a major supplier of oil and gas for itself and for Europe.

The North Sea fields, large though they were, did not rival the scale of those found in places such as Texas, Russia, or the Middle East. In 1999, oil production from the United Kingdom's portion of the North Sea peaked; gas production peaked the following year. Since then, the output of both fuels has fallen by more than half. Nonetheless, the North Sea remains the largest single provider of hydrocarbons for the United Kingdom, generating billions of dollars each year for the government and private companies.

While pumping massive quantities of hydrocarbons from beneath the North Sea, the United Kingdom—along with the rest of Europe—was also developing and implementing the world's most comprehensive and ambitious climate-change policy. In keeping with the Kyoto Protocol, an internationally negotiated climate treaty signed in 1997 and that entered into force in 2005, the European Union implemented an economy-wide cap-and-trade program.[17] In addition, the United Kingdom and other governments heavily taxed fossil fuels, including gasoline and diesel; spent large sums to subsidize renewable sources such as offshore wind; and took other steps to reduce greenhouse-gas emissions. While scholars have debated the effectiveness of this particular mix of policies,[18] it is clear that the United Kingdom and its former colleagues in the European Union have been at the forefront of the effort to implement climate-change policies for most of the past decade.

As the United Kingdom has moved toward renewables and a lower-carbon future, and as its North Sea fields have lost their luster, a new potential source of hydrocarbons has emerged. This time the potential does not lie offshore but instead beneath the hedgerows of places such as Yorkshire, where shale gas has begun to attract the attention of Britain's government and of oil and gas companies from around the world.

The United Kingdom has an estimated 26 trillion cubic feet (Tcf) of natural gas trapped in shale. While not of the same scale as the prospects in places such as Argentina (~800 Tcf) or China (~1,100 Tcf), 26 Tcf is nothing to sneeze at. If 10 percent of these estimated resources could be recovered (a reasonable guess for shale formations), the value of that natural gas would be $13 billion dollars, assuming a modest gas price of $5 per thousand cubic feet.

And although the British government has remained committed to achieving its long-term climate goals, the prospect of a low-cost natural gas supply located within its own borders holds substantial appeal. Even under ambitious climate policies, natural gas will continue to provide an important part of the British energy mix for decades to come, and a domestic supply source would be far preferable to relying on imports.

In 2015, the United Kingdom imported more than half of the natural gas it consumed, including a substantial amount of LNG. This reliance on costly imports, coupled with strong demand for gas, has had a major effect: the average price for natural gas in the United Kingdom grew from roughly equal to the U.S. price in 2005 (before the shale revolution) to more than twice the U.S. price in recent years.[19] Partly as a result, electricity prices in England (like most of the rest of Europe) are roughly twice those of the United States.

Despite these higher prices and the country's climate goals, natural gas consumption has grown steadily, providing about 40 percent of the United Kingdom's electricity in early 2016 (renewables, nuclear, and coal each provided about 20 percent).[20] Put simply, shale gas has the potential to help lower the high cost of energy in the United Kingdom.

Responding to this fact, Britain's government has largely supported shale gas development within the country. A guidance document issued by the government in 2017 describes how domestic shale could displace natural gas imports and makes the case that the environmental risks of development are generally modest and manageable.[21]

But many in Britain's environmental community see the government's support for fracking as a betrayal of its climate goals and aren't ready to accede without a fight. Demonstrators in England have staged a range of bold protests to block industry activity, going well beyond even the most boisterous antifracking groups in the United States. For example, campaigners in 2013 set up an encampment near a roadside about thirty miles south of London, hoping to prevent a British company from drilling. At its peak, the protest attracted more than one thousand campers, who at times blocked the path of trucks trying to make their way to the site.[22]

Other proposed drilling locations around the United Kingdom have seen similarly vociferous resistance. In the small village of Upton, near Liverpool, dozens of demonstrators camped out for more than eighteen months to prevent the development of coalbed methane in the region, fortifying their position with towers, tunnels, and even a makeshift moat. One demonstrator stated: "we can look to America for 10 years' worth of evidence as to what fracking means for communities. We won't have our health and environment ruined just to make a small number of people a large amount of money."[23]

Local opposition isn't the only reason fracking has moved slowly in the United Kingdom. The depth of the shale plays that most interest companies are deeper than many U.S. shale formations—on the order of 9,000 feet below the surface—making it more expensive to drill and frack a well. In addition, stricter regulations apply in the United Kingdom than in most U.S. states. For example, companies are required to monitor groundwater near the well site for one year prior to fracking, add additional rings of steel well casing, and follow more restrictive local planning guidelines.[24]

For these reasons, no large-scale United Kingdom shale gas production was taking place as of early 2017 (though there have been some test wells drilled), and projections of future production are highly uncertain. At the low end, one study projects shale gas production in 2030 of about 35 billion cubic feet (bcf) per year, about 1 percent of domestic consumption in 2014. At the high end, another study projects production of roughly 1,375 bcf by 2040, almost 60 percent of 2014 consumption.[25]

So are the protestors right? Will the United Kingdom frack its way into climate disaster, or could shale gas actually help the nation meet its international climate commitments? A 2016 report prepared by the Committee on Climate Change, an independent research body established by the UK government to offer advice on climate issues, provides a thoughtful answer to both questions: maybe.[26]

As in the United States, where increased natural gas use has the *potential* to help the country meet ambitious climate goals, the committee found that developing shale gas can help the United Kingdom meet its climate targets only under certain conditions. First, methane (and other GHG) emissions from shale development would have to be strictly monitored and kept to a minimum. Second, the new supply of shale gas would need to displace natural gas imports, rather than increase the overall amount of natural gas consumed. And third, any increased emissions associated with shale production would need to be met by emissions reductions elsewhere in the economy, ensuring that the United Kingdom remain within its carbon budget (the maximum level of emissions the nation can emit and stay within its climate goals).

Clearly, two major questions remain. First, will the political, economic, and geological stars align to allow the United Kingdom to develop shale at large scale? And second, if those stars align, will the government abide by the guidelines laid out by the Committee on Climate Change? The answer to each of these questions remains to be seen.

VACA MUERTA, ARGENTINA

Vaca muerta, in Spanish, translates literally to "dead cow." And on a 2016 trip to Buenos Aires, I became familiar with two distinct applications of this phrase. First, Argentinians are well known for their healthy appetite for beef. Over the span of four days, I consumed something on the order of five pounds of steak.

Second was the real reason for my visit: the other Vaca Muerta, one of the most promising shale gas resources in the world, stretching hundreds of miles through the western half of Argentina into Patagonia. As part of a technical-assistance program led by the U.S. State Department (more on that later in this chapter), I traveled with academics, regulators, and other experts to Buenos Aires to make presentations to the Argentinian government about what lessons we'd learned in the United States over the past ten or so years about regulating and managing the growth of fracking.

This was a time of transition for Argentina. A new president, Mauricio Macri, had been elected in early 2016 and now governed from the Casa Rosada, Argentina's pink-hued equivalent of the White House. President Macri promised to offer a more market-friendly atmosphere than that of his predecessor, Cristina Fernández de Kirchner, and this included attracting investment in the Vaca Muerta. (President Kirchner had discouraged oil companies from coming to Argentina by, among other things, seizing the assets of the Spanish oil company Repsol in 2012 after accusing it of underinvesting in the country.)

For decades, Argentina has been a leading South American producer of oil and gas, easily producing more than it has consumed. However, steadily growing consumption and weak investment in developing new supplies led Argentina to become a net importer of both oil and natural gas by the early 2010s (see figures 10.3 and 10.4).

The shale era offered an opportunity to reverse Argentina's slide in production, potentially making it a net exporter again. Argentina possesses more shale gas resources than almost any other country (recall figure 10.1), and the biggest prize is the Vaca Muerta.

To encourage companies to invest in the Vaca Muerta—and to ease some of the concerns raised by the appropriation of Repsol's assets—the Kirchner administration offered artificially high prices for oil and gas produced in Argentina. While this strategy has the potential to strain government budgets, it has succeeded in attracting interest from some of the world's biggest companies. Chevron partnered with the Argentinian oil company YPF to drill hundreds of wells in the region, and ExxonMobil, according to press reports, is considering an investment of more than $10 billion in the Vaca Muerta.[27] Other major companies, such as Dow Chemical and France's Total, are also interested in the region.

But developing the Vaca Muerta is not a simple matter. The region where most drilling has occurred to date, called the Añelo department in

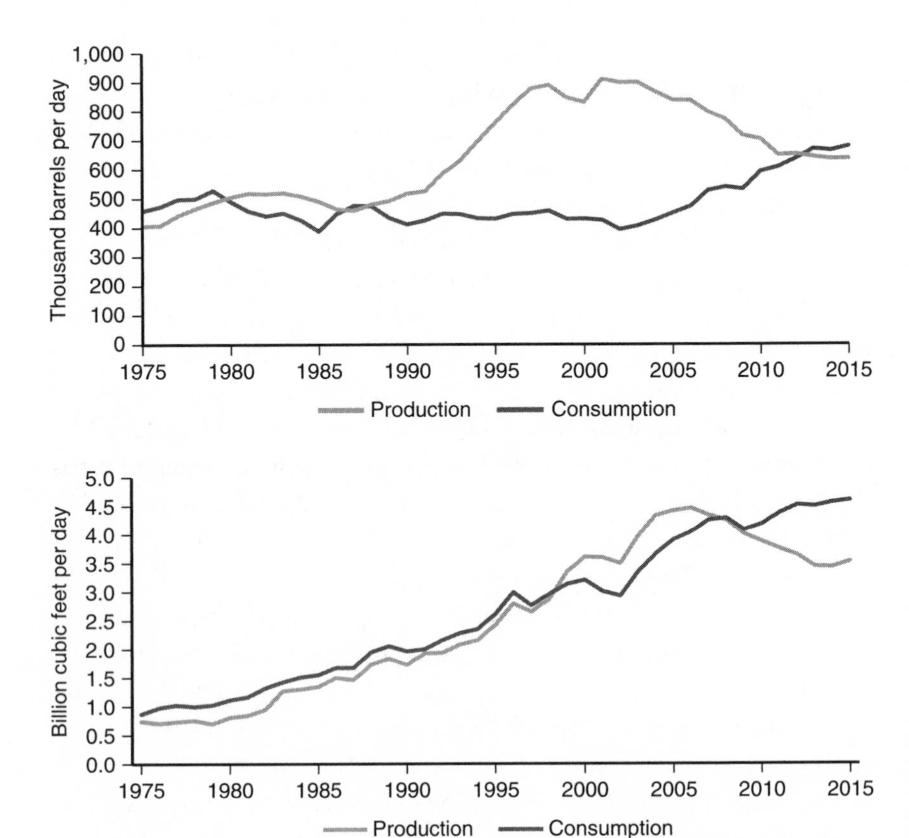

FIGURES 10.3 AND 10.4 **Argentina oil (10.3) and gas (10.4) production and consumption trends**

Source: BP, Statistical Review of World Energy 2016, http://www.bp.com/en/global/corporate/energy-economics/statistical-review-of-world-energy.html.

Neuquén province, is extremely rural. Far more rural than even the Bakken in North Dakota, the Añelo region would certainly face challenges absorbing the influx of people, vehicles, and equipment associated with a ten-billion-dollar drilling campaign. Even the limited drilling activity that occurred through 2016 strained Neuquén province, as there are few local suppliers of equipment or services needed to develop a major new field. For example, all of the sand used for proppant during fracking must be trucked in from the coast, much of it imported from Brazil or the United States, carried on roads designed to accommodate local farmers and herders, not thousands of heavily laden eighteen-wheelers.

However, other aspects of the Vaca Muerta bode better for development. Smaller-scale oil and gas production has occurred for decades in the region, so there is already a substantial network of pipelines in place to move the oil and gas to eager consumers. Similarly, the provincial officials we met with in Buenos Aires seemed quite knowledgeable about regulating oil and gas activity and appeared to be well on their way to developing a thorough set of regulations to minimize the environmental risks associated with shale development.

More broadly, the oil and gas industry is no stranger to developing major fields in remote locations. And when the prize is big enough, it has proven its ability to tackle greater challenges than a rural arid region of western Argentina—think of Alaska's frozen North Slope or the harsh North Sea. Remote locations can be overcome if the right conditions are in place. In short, it looks as though the Vaca Muerta won't be dead for long.

SELLING FRACKING TO THE WORLD

While the 2016 U.S. presidential campaign will be remembered for the rise of businessman-turned-president Donald Trump, the Democratic primary race between Hillary Clinton and Bernie Sanders highlighted a fundamental difference between two approaches for dealing with energy policy in the context of a changing climate. While the Clinton campaign struck a similar tone to that of the Obama administration's active yet relatively pragmatic actions on climate change, Sanders argued for more radical steps to upend the nation's, and the world's, energy system.

Among the differences highlighted during the campaign were the role of nuclear power, views on the "keep-it-in-the-ground" movement (which seeks to halt as much fossil fuel production as possible), and—you guessed it—the discussion of whether to ban fracking across the United States. The debate also strayed into international waters, as allies of the Sanders campaign accused Clinton of actively promoting fracking around the world during her tenure as secretary of state.

And though the language used by Clinton's critics often overstated the case, they were—in a certain sense—correct. While news articles with titles like "How Hillary Clinton's State Department Sold Fracking to the World"[28] did little to bring nuance to the discussion, the basic fact remains that the U.S. government has offered (and continues to offer) assistance to other countries interested in developing their own shale resources via fracking.

Critics of this policy weave a narrative that casts Secretary Clinton in the role of drug dealer, selling fracking to nations unaware of its dangers. However, many nations have envied the U.S. resurgence as an oil and gas powerhouse and are eager to learn what fracking could do for them. These nations ranged from Poland to Indonesia and from Algeria to Argentina, each with their own set of geopolitical, environmental, and economic interests in learning not just how to frack but how to regulate and manage the industry if and when it reaches their shores.

Of course, the United States was also advancing its own interests by encouraging certain nations to become less reliant on energy from traditional geopolitical adversaries such as Russia and Iran. In a 2012 speech delivered at Georgetown University, Secretary Clinton clearly expressed the strategic importance to the United States of preventing allies from becoming solely dependent on nations like Russia, which has, in years past, limited natural gas supplies to Europe to exert political influence. The importance of shale gas development in Europe has not been lost on Russia. U.S. intelligence and press reports suggest that Gazprom, the Russian state-controlled natural gas supplier to Europe, may have helped finance antifracking movements in Eastern Europe, and a perusal of fracking coverage in *RT* (formerly *Russia Today*), a Kremlin-financed cable channel, showcases a litany of fracking-related horrors.[29]

Clearly referring to this issue, Secretary Clinton said, "Energy monopolies create risks. Anywhere in the world, when one nation is overly dependent on another for its energy, that can jeopardize its political and economic independence."[30] While she did not name Russia explicitly in this passage, my guess is that she wasn't referring to the United States and its heavy reliance on imported Canadian crude oil.

In keeping with these priorities, the U.S. State Department announced in 2010 the Global Shale Gas Initiative, which brought together seventeen countries from around the world with experts from academia, government, industry, and NGOs from the United States to discuss the potential for shale gas development in their own countries.[31] And while supporters of Bernie Sanders's presidential campaign might see this type of event as a forum for U.S. companies to advance their own interests, the bulk of activities carried out under the program have been to share regulatory expertise developed in the United States over the past decade.

And how would I know?

In June 2016, I participated in one of these workshops in Buenos Aires. Now rebranded with a more bureaucratic moniker and led jointly by the State Department and the Department of the Interior, the Unconventional Gas Technical Engagement Program (UGTEP—really rolls off the tongue, doesn't it?) sends experts from governments, academia, and other regulatory organizations to countries like Argentina, China, and Indonesia. At those meetings, the U.S. delegation shares best practices with government officials looking to develop their own regulations on shale and other "unconventional" oil and gas resources.

During the two-day workshop, it was clear that Argentina wanted to develop the Vaca Muerta. We were there to help them understand what we'd learned and to describe ongoing challenges, such as methane emissions and wastewater disposal. For my portion of the event, I described the challenges faced by some local governments experiencing rapid population growth due to the shale revolution, and I talked about how setting up the right kinds of tax policies could help reduce some of those problems.

Other experts talked about regulating all of the major issues discussed in this book: preventing damage to drinking water, reducing harmful air emissions, preventing earthquakes, and more. If this workshop was designed to "sell fracking to the world," I'm pretty sure we weren't the "A-Team."

SUMMING UP

Shale development is spreading around the world, but slowly. While a limited number of nations have imposed moratoriums or bans, a larger number have sought to learn about fracking and apply it in their own backyards. U.S. companies have invested abroad to access the shale, and foreign companies have invested in the United States to learn how fracking works. Some countries, namely, China and Argentina, are already producing gas and oil from shale, with an eye to scale up in the coming decades.

However, most international shale development has moved fairly slowly due to a variety of factors. Perhaps most importantly, landowners in every country other than the United States do not receive a share of the revenues from oil and gas produced beneath their land, reducing their incentives to welcome new oil and gas drilling. At the same time, local protest movements driven by fears of environmental degradation have slowed fracking's spread.

11

DO PEOPLE LIVING NEAR FRACKING LOVE IT OR HATE IT?

Over coffee one morning in 2014, I had a long conversation with a sixty-something farmer named Martin, who lives in the tiny town of Zahl, in Williams County, North Dakota. Zahl sits about thirty miles north of Williston, which, with a pre-Bakken boom population of about 12,000, counted as the region's "big" city. Martin, who had served multiple terms as a county commissioner, told me about how things had changed over last ten years.

In 2005, things were quiet. He loved the peace of the place, sitting on the porch in the summer and watching the stars meander across the prairie sky. He recognized just about all of the pickup trucks that drove through town. But his kids had moved away from the region a long time ago, and there were few opportunities for those that remained. The population was slowly shrinking, and it wasn't at all clear whether western North Dakota had much of a future.

Then things began to change. Stories starting circulating about a rock formation called the Bakken, which was going to spark a new era of North Dakota oil production. Like many of his neighbors and other government officials, Martin was skeptical about the hype. There had been oil booms before in North Dakota—the last one was in the late 1970s and early 1980s—and they usually came and went without much long-term impact. But landmen soon started knocking on doors and offering thousands of dollars per acre for the rights to drill underneath local farms. Martin saw more trucks on the road—not the tractors and cow pullers that had always roamed the narrow dirt paths, but big, hulking, dangerous eighteen-wheel petroleum tanks.

The quiet of the place had gone. Sitting on the porch at night, he could hardly make out the stars for all the bright orange flares lighting up the prairie (though, to be fair, he also hated the site of wind turbines on the horizon). There were heartening new stories about kids returning to western North Dakota: daughters moving back to start a restaurant, sons making $100,000 per year driving a truck. This was great news. But things were different now. More hectic. More traffic. Constant construction. As our conversation ended, Martin told me he was thinking about moving somewhere else. Somewhere with a little more peace and quiet.[1]

A couple days later, after a long day of travel and interviews, I struck up a conversation at the hotel bar in Williston with a welder from Colorado. Wiry, well tanned, and somewhere in his forties, he'd been employed by a major international oil company for the past ten years or so, working jobs from Alaska to Nigeria to Malaysia. He made a good living, but the work was hard, and he was sick of the travel. Then they asked him to go work in North Dakota, in the Bakken.

These days, he could fly home every weekend to see his wife and daughter. He got an $80,000 bonus and a promotion. He got a nice new truck. For the first time, he was feeling confident about the balance in his daughter's college fund. And he was pumping oil and gas in the good old U.S. of A. He described watching and worrying over the occasional spills and well blowouts, but for the most part, he was proud of his job and proud of his industry.

These two types of stories are common in the oilfield. While pro- and antifracking advocates often paint one-sided pictures of the boom, highlighting only the upside or only the downside, ambivalence and ambiguity are far more common in places like Williston. I've heard variations on these two stories in pretty much every setting you can think of: across a conference-room table, on a neighboring barstool, at a sushi restaurant, in the basement of a grain elevator, at a McDonald's, at the local newspaper. Everyone has their own experience, and the stories I've heard are rarely simple.

On the whole, the balance of conversations I've had lean toward the positive side of the ledger. Although I've met many people frustrated by the traffic, the noise, or the faster way of life, I've met surprisingly few who have experienced environmental degradation or described it happening to their neighbors. A larger portion of the stories paint a picture of people grateful for the economic opportunities afforded to them, their families, and their communities.

The rest of this chapter tells a few of those stories—some based on specific individuals, others based on composites—of the people I've met across America living through the shale revolution.

YES, IN MY BACKYARD

Fairview

I met the mayor of Fairview, Montana (population ~900), at the car-repair shop he runs at the town's main intersection. Fairview lies just across the border from North Dakota, about an hour's drive from Williston and the heart of the Bakken boom. The mayor brought along his son, who looked to be about ten years old, and we wandered across the street to grab lunch at one of the town's two restaurants, a small clapboard diner with a dusty parking lot.

The lot was packed with a mix of old-timers and newcomers. Old, almost classic, trucks with bulbous hoods were lined up next to nearly new eight-foot-tall F-150s coated in heavy layers of dust. Inside, farmers in their sixties wearing overalls and mesh baseball caps sat next to young oilfield hands with mud-caked boots and flame-retardant outerwear. The diner was woefully understaffed, with two servers scurrying from table to table, taking two orders for every one they delivered.

As we waited for our BLTs, the mayor and I talked about what it was like living in Fairview these days. He pointed outside at the eighteen-wheel trucks that passed by every few seconds. They were a mixed fleet: sand trucks, towing two rust-colored tubs of sand to be used as proppant, and liquids haulers, towing one shiny silver tube that could hold fresh water on its way to a frack job, wastewater on its way to a disposal well, crude oil on the way to a rail terminal, or any number of other oilfield fluids. He talked about how the property values in Fairview had just about doubled over the past few years and how the trucks were tearing up the roads.

But the thing I remembered most was the mayor's description of economic opportunity. Before the Bakken, there was little hope that his son, sitting beside him and fully absorbed in his french fries, would stick around Fairview after high school. Farming didn't much interest the younger generation, and Fairview's two restaurants, one bar, and the family repair shop didn't hold much promise in the way of excitement or employment. But, he said, the Bakken has been luring back lots of young people in recent years. Several friends of his had sons and daughters that had recently moved back to the area, driving trucks, working on rigs, one even making good money playing music. He didn't know how long it would last, but for the first time, he had some hope that his son might stay near Fairview to raise his family.

Dimock

Dimock Township, Pennsylvania (population ~1,500), is often cited as a casualty of the fracking boom. Featured in everything from the anti-fracking film *Gasland* to an op-ed by Mark Ruffalo, Dimock is the go-to example of what can go wrong when fracking comes to town. As I describe in chapter 3, these concerns are not without reason. Shale gas wells drilled in the area by one company contaminated drinking water for more than a dozen homes, leading to large fines, legal settlements, and an order from the Pennsylvania Department of Environmental Protection that barred the company from drilling any more wells in the township.[2] As a result, when I visited in the fall of 2013, no new wells had been drilled in Dimock for several years.

Press coverage has described the split between residents who oppose drilling and those who support it,[3] and when I drove into town to meet with several local leaders, I was curious to hear their perspective. As it turns out, local elected officials (called township supervisors) had been trying to get drilling going again in Dimock for years. While the supervisors acknowledged the contamination problems for the roughly twenty families along Carter Road, underneath which a failure to cement gas wells properly led to stray gas, they described how hundreds of other residents were missing out on the opportunity for royalties from natural gas production.

Because of the hold on new drilling, landowners who had expected a sizeable influx of cash were eager for the state to lift the moratorium, ready to take advantage of the Marcellus shale that lay beneath. The township supervisors were exasperated with the state for slowing things down and with the small number of town residents who continued to fight the gas company through the press and in a federal lawsuit (in 2016, several years after my visit, a federal jury awarded $4.2 million in damages to the families that sued,[4] though this award was subsequently overturned and a new trial ordered).

Instead of finding a town weary of fracking and its consequences, as I had expected, I found one that was doing all it could to bring the rigs back. Of course, not all of Dimock's residents supported new gas drilling, but judging from their elected leaders, a majority certainly did.

Coalgate

The city manager of Coalgate, Oklahoma (population ~1,900), couldn't stop talking about the new fire station. Coalgate sits in the southeastern corner of Oklahoma, about two hours south of Oklahoma City and two hours north of Dallas. It's a quiet town, especially since the early 2000s, when the city's leading employer, a jeans manufacturer, shut its doors and laid off hundreds.[5] Without this employment base, the city had begun to erode. Longtime residents were moving out, high-school graduates were moving away, and it wasn't clear how long the city would survive.

But around the mid-2000s, drilling rigs started popping up on the outskirts of town and across Coal County, bringing with them hope of a reprieve from a depressed economy. The Woodford shale, which after the Barnett was one of the first shale gas fields in the world to be developed at a large scale, attracted dozens of companies and provided a major boost to the local economy. This activity increased tax revenues, helping Coalgate build a new fire station and EMS dispatch center. It also allowed the city to provide 24-7 emergency services to the community for the first time.

Drilling in this part of the Woodford has since slowed substantially, and when I visited in 2014, there were just two rigs operating in the area.

But in 2010, a rifle manufacturer moved into the old jeans plant, bringing a more stable employment base.[6] As it turns out, the shale boom was something of an economic bridge for Coalgate, stabilizing the community when its manufacturing-based economy remained uncertain. Along the way, the infusion of tax funds helped it provide better services to its citizens. As in many other oil- and gas-producing regions, a drilling boom wasn't an economic panacea, but it certainly helped the community outlast a downturn.

NOT IN MY BACKYARD

Williston

For all the economic growth and prosperity that the Bakken has brought to western North Dakota and eastern Montana, it's been accompanied by major changes to the way of life for longtime residents. One morning in 2015, I sat down for eggs and sausage across the table from a burly man in his fifties or sixties with a scraggly white beard, dressed in overalls and a camouflage mesh ballcap. A wheat farmer north of Williston, he described how much more it cost to do just about everything since the rigs came to town: buy food, hire employees, send his crops to market, and more.

As I described in chapter 9, when an economic boom comes to town, it does not lift all boats. For longtime residents like my breakfast companion, whose kids moved away years ago and whose business does not benefit from the Bakken boom, there wasn't much upside to living in "Boomtown, U.S.A.," the not-undeserved nickname coined by Williston's visitor's center. He had leased his land early in the drilling process, before prices went through the roof. A pipeline carrying produced water from nearby wells had ruptured near his property, damaging a neighbor's wheat fields potentially beyond repair (research by one of my former colleagues at Duke explores this issue across the region).[7]

Crime had also increased dramatically. While rates of violent crime were falling on average across the United States for most of the past decade, in Williams County they had grown by almost 500 percent.[8] And the day-to-day hassles were everywhere. The shelves at Wal-Mart were sometimes stripped bare of necessities like toilet paper and toothpaste. Picking up a drive-thru breakfast had transformed from a quick stop into a twenty-five-minute ordeal, as lines of huge white pickup trucks idled, ten or more at a time, waiting for the chance to order a lukewarm sausage sandwich. From his porch at night, it sometimes seemed like the world was on fire, as bright orange flares cast an ominous glow over the prairie. And the eighteen-wheelers never stopped.

Anthony

Anthony, Kansas, is a sleepy place. When I visited in late 2014 and went searching for lunch on Main Street, there was just one open restaurant sitting beside a row of empty and abandoned storefronts. The marquee on the dilapidated movie theater looked like it hadn't been cleaned since the 1950s, and the only evidence of its continued functioning was a tattered poster advertising a single weekend showing of an old musical. Anthony is in south-central Kansas, just six miles north of the Oklahoma border, and as the seat of Harper County, it boasts a courthouse dating to the mid-1800s. In the early 2010s, dozens of landmen started appearing at the courthouse to research land ownership. Soon after, they were joined by drilling rigs coming to tap the Mississippian Lime formation that lay several thousand feet below the surface.

The courthouse, made from red brick with a grey stone base, was where I met the county clerk, who walked me through the government financial documents that had prompted my visit. Along the way, we got to talking about the series of small earthquakes that had shaken her and the rest of the community (see chapter 5). She mentioned that a building inspector had recently visited and expressed concerns about the foundational integrity of the courthouse. She described the cracks that a number of neighbors had recently found in their walls, which appeared to be caused by the small but frequent temblors.

The earthquakes weren't ruining people's lives or causing widespread harm, but it was clear that something wasn't right. The county had no desire to abandon its treasured courthouse, but the government also had no desire to spend millions of dollars resecuring the foundation. There was a lot of economic good happening around Anthony because of the drilling boom, but the earthquakes were starting to make the place feel eerie, and people were worried. When was the shaking going to stop?

Center Township

The morning after the unexpected election of Donald Trump as president of the United States, I drove to a meeting with two township officials in Center Township, Greene County, in the southwestern corner of Pennsylvania. Greene County has a long history of mining, shipping, and burning coal. During my first visit in 2013, one local official gave me a glossy booklet highlighting that year's "King Coal Show," which, among other festivities, crowned one lucky high schooler with the title "The 2012 Pennsylvania Bituminous Coal Queen" (bituminous refers to the type of coal produced in the region). However, recent years have seen some of the county's largest mines close, along with a large coal-fired power plant, while Marcellus shale development has grown rapidly.

Back in Center Township, after a twenty-minute conversation in the small mobile home that served as headquarters, we piled into a high-riding F-150 so the commissioners could show me around the township. Bumping along the undulating dirt roads in the lushly forested countryside, my tour guides started by recounting their surprise and joy when waking up that morning and hearing of Trump's victory.

A few minutes later, as we passed several new well pads sporting five to ten natural gas wells, the mood turned sour. Pipelines, my guides explained, had scarred the hills, crisscrossing the township and cutting through the forests. We parked near the top of one peak, then got out to take in the surrounding landscape. On one side of the hill, where little shale development had occurred, dense networks of trees dotted with homes and farms filled the view. On the other side, well pads had cleared many of the neighboring hilltops, and long lines of bare ground cut through the thick forest. I asked my guides if they thought the economic boost was worth these changes to the countryside. Maybe in the short term, they answered. But the wells and pipelines would be there for decades, maybe the rest of their lives. The thought made them cringe.

NOT IN THE PLACE I LOVE

As oil and gas development has moved into new corners of the United States, communities have responded in a variety of ways. Some are enthusiastic; others have been skeptical. These divergent reactions have been driven in part by each city, county, or state's familiarity with the oil and gas industry. In parts of Texas, Oklahoma, New Mexico, and other regions where drilling rigs, pumpjacks, and pipelines have long been a mainstay of the local landscape, there tends to be far less concern over, and opposition to, fracking and all of the issues that come with it. In other parts of the country, where the sights and sounds of the industry are new, locals understandably tend to have more questions and in many cases more concerns.

But in some cases, this generalization doesn't hold up as neatly as one might expect. In late 2016, I participated in the first in a series of dialogues on shale development at the Aspen Institute, a nonprofit organization dedicated to bringing people together to develop leadership goals on complex topics like oil and gas production. This dialogue gathered experts from regulatory agencies, oil and gas companies, environmental-advocacy groups, academia, and other research institutions for several days of discussions on how best to govern oil and gas development in the shale era.

On the second afternoon of the dialogue, as the group discussed how the concerns of local communities vary from place to place, Rick McCurdy, an expert on hydraulic fracturing with a large independent oil and gas company, shared a story.[9] Rick described how, as a part of the oil and gas industry for decades, he and his wife, Janice, had moved from place to place: Texas, Oklahoma, and Alaska. They were steadfast supporters of the industry, confident that the work they did provided the essential lifeblood—energy—that underpinned a healthy domestic and global economy. They had also seen firsthand the economic benefits that oil and gas production could bring to the communities in which they'd lived.

At the same time, they were confident that—by and large—the industry did a good job of preventing environmental damage in the regions where it operated. Given the extraordinary technical challenges of drawing highly combustible and hazardous products from miles beneath the earth's surface, often under inhospitable conditions, they felt that most companies did a good job of controlling the hundreds of variables that could turn an ordinary well into an extraordinary environmental problem.

Rick and Janice had largely scratched their heads when observing opposition to fracking in parts of Pennsylvania, New York, or Colorado, where drilling rigs were an unfamiliar sight.[10] Much of the opposition, Rick hypothesized, rested upon a lack of familiarity with the industry. These folks didn't understand the economic benefits that the industry could bring to their region, and they didn't understand that the risks of drilling were actually pretty small.

But in September 2016, a large independent oil company called Apache announced an astonishing new find. They called it Alpine High. Located in western Texas's Delaware basin, where the industry has been a mainstay of the local economy for decades and where few thought there was any more oil left to be discovered, Apache announced that Alpine High likely held more than 3 billion barrels of oil and 75 trillion cubic feet of gas.[11] If Apache were able to extract just 10 percent of this oil and gas, it could bring in revenues of almost $40 billion (assuming oil prices of $50/bbl and gas prices of $3/thousand cubic feet). In short, this was a big deal and could mean decades of drilling ahead for Alpine High.

As Rick learned more about Apache's discovery, he was amazed at the scale of the find and the fact that companies were still turning up new plays in picked-over regions like West Texas. It was something of a gold-star moment for the industry—even with oil prices stuck in the doldrums and gas prices near historic lows, companies were still pushing forward, still finding opportunities where previous generations had only seen scrubby hills.

But Rick's wife, Janice, wasn't sensing that same opportunity, nor was she feeling the same pride as her husband. Instead, her feelings were of fear and apprehension. Janice had spent much of her childhood enjoying the natural beauty of the Davis Mountains, the West Texas hills that sat atop much of the Alpine High find. When she imagined oil rigs crawling into those hills, brush being cleared to make way for new pipelines, and well pads dotting the landscape, she balked at the notion of oil and gas development encroaching into her childhood refuge, a place she still loved to hike and explore. She worried about risks to the groundwater and the springs of Balmorhea State Park, a natural oasis in the West Texas desert.

As Rick told his story to the group, it became obvious that even for those who try to make disinterested, logical, and objective calculations about the benefits and risks of fracking, the arrival of drilling rigs and fracking crews into a place that you hold dear changes the equation. Fear of even remote risks can overwhelm long-held beliefs and trump the most rational cost-benefit analysis. And even if there are no hazardous spills, no methane migration, and no harmful air emissions, the mere presence of well pads, pipelines, and heavy trucks can fundamentally change the character of a beloved rural region.

While Rick believed that the industry could safely develop Alpine High, Janice wasn't so sure. Even for ardent supporters of the oil and gas industry, when fracking (or any other potentially disruptive activity) comes to a place that you love, support might no longer be a given.

A LOVE/HATE RELATIONSHIP

For the vast majority of people I met in the oil and gas fields of America, fracking has been a mixed blessing but a blessing nonetheless. It has created enormous one-time injections of wealth into struggling communities, but the ups and downs of the energy industry mean that the long-term economic impacts are far from certain. It has kept communities together by providing job opportunities for the next generation while at the same time increasing tension within communities as thousands of outsiders pour in and neighbors argue about the pros and cons of the industry. It has created a sense of hope for the future of rural regions, all while raising new concerns about long-term environmental and economic risks. For the bulk of the people I've met over the years, the benefits have outweighed the concerns. A common sentiment ran along the lines of: "I love having new opportunity for me and my family, but I'm a little wary about the future, and I hate the traffic."

Despite the widely touted notion that fracking leads inevitably to widespread environmental harm, most of the concerns that people expressed to me were not about the environment. Among the hundreds of government officials I interviewed and the hundreds of people that I met at restaurants, bars, libraries, and hotels, the number who had personally experienced any type of contamination was precisely zero. But several had neighbors whose water had been infiltrated by stray gas, and a more substantial number told me the stories of friends who lived near well blowouts, spills, and other events with real environmental consequences. Others described changes in the physical landscape, as well pads, compressor stations, and other infrastructure spread across the hills of southwestern Pennsylvania or the prairie of western North Dakota.

Of course, sentiments varied from place to place. In parts of Pennsylvania and Ohio, where oil and gas drilling at this scale was new to everyone, the impacts of trucks on the road, flares at night, and a spike in housing costs came to the fore. These communities, which are more densely populated than most of the oilfields farther west, saw lots of benefits. But those benefits were tempered by the fact that much of the work was going to outsiders driving into town from Texas or Oklahoma, creating a sense of resentment among some of the folks I talked to across the diner counter or the barstool.

In most of the longtime producing regions like western Texas, central Oklahoma, central California, and elsewhere, this was all old hat. Almost everyone I met welcomed the boom, and several offered variations on an old oilfield bumper sticker: "Dear God, give me just one more oil boom. I promise not to piss it away this time." The people who live in these longtime producing regions know about the environmental risks and mundane hassles that come with the territory, but they have by and large accepted the trade-offs involved.

Parts of Wyoming, North Dakota, and Montana have all seen oil and gas production for decades, but mostly on a smaller scale from what they'd experienced in the shale era. And because the economies in most of these regions hadn't been structured around oil and gas production, they've experienced some of the most acute booms. Plenty of locals wondered aloud how long it could last. By 2012, rural western regions of Wyoming and Colorado focused on natural gas had already experienced a large boom and bust, peaking around 2008 and slowing dramatically by 2012 as gas prices crashed (see chapter 9). Cities like Pinedale, Wyoming, and Rifle, Colorado, had expanded rapidly to accommodate the new industry, then watched in trepidation as the new hotels and restaurants slowly emptied. Regions like North Dakota's Bakken and southern Texas's Eagle Ford, where the boom has been mostly about oil, followed a similar pattern, with activity peaking in 2014, followed by a downturn a few months later as oil prices swooned.

When I returned home from the oilfield to college towns like Durham, North Carolina, or Ann Arbor, Michigan, where little or no oil and gas production takes place, I'm often struck by the more visceral attitude that many take toward the industry. Strolling downtown for dinner or stopping by the farmer's market to pick up produce, I have frequently been stopped and asked to sign petitions to "Ban Fracking!" in North Carolina or Michigan. Walking or driving through my neighborhood, I'll see "Ban Fracking!" yard signs. Wandering across campus during campaign season, I've come across rallies where young and old alike carry signs that proclaim "Yes We Can (Ban Fracking!)."

But when I travel to Williston, North Dakota; Midland, Texas; Vernal, Utah; Bakersfield, California; Anchorage, Alaska; or any other place where the oil and gas industry plays an outsized role in the daily life of residents, I've never seen a "Ban Fracking" sign (with one exception: along Carter Road in Dimock), and I've never been asked to sign a "Ban Fracking" petition.

This difference in attitudes may not be surprising, but it illuminates a key point: while the shale revolution has caused problems that range from the mundane to the severe, for most families living in the affected areas, the good tends to outweigh the bad. From the hundreds of conversations I've had, most people living in communities where oil and gas production underpins the economy experience the risks and rewards of the industry in real time. Some have signed leasing bonuses making them instant millionaires, and others have had their water affected by stray gas. Many have neighbors and friends whose children have moved back to town to drive a big rig, but many also know someone who has been injured or killed in traffic accidents caused by those same eighteen-wheelers. The local diner is busy and profitable like never before, which means more wealth coming into the community but also a thirty-minute wait to get a sandwich.

And while research in this area is limited, some related work seems to back up my hunch that people living closer to producing regions tend to show more support for fracking. For example, a 2012 study showed that people living in Pennsylvania, where Marcellus development was ongoing, were more likely to support the industry than people living just across the border in New York, where a fracking moratorium held sway.[12] Another found that residents of a heavily drilled Pennsylvania county were more supportive than a neighboring one where far less drilling had occurred.[13] A 2016 study found that people living in places where the oil and gas industry supports the local economy are more likely to support fracking.[14] One study from Pennsylvania showed that public support for fracking was stronger in regions where revenues from oil and gas development were benefiting local governments and helping provide public services.[15] Others have found that residents living in oil- and gas-producing states tend to support the industry more than those who live elsewhere, even after disasters like the explosion and oil spill of the Deepwater Horizon.[16]

Along similar lines, a 2016 study found that the farther away people lived from oil- and gas-producing regions, the more likely their opinions were to be shaped by simple political affiliation, with Republicans tending to support fracking and Democrats tending to oppose the industry.[17]

In short, people living in the oilfield have formed their own opinions about the pros and cons of fracking and the industry writ large. While everyone is entitled to their opinion about the benefits and costs of fracking (regardless of whether they live in a region rich with oil and gas production), it seems to me there is some reason to give additional weight to the people who live in the communities that are directly affected, for better or for worse.

SUMMING UP

Traveling through small towns and rural counties, I've met hundreds of people living through the shale revolution. I've heard scores of stories describing how the shale boom restored a region that was in decline, providing new economic opportunity and encouraging young people to move back home. I've heard dozens of stories lamenting the truck traffic, increased crime, and the sense that a small town's character has been changed by an influx of new workers and a new industry. On the whole, the positive tends to outweigh the negative, but the stories of the people living through the booms and busts are rarely simple.

12

WHAT'S NEXT?

Despite the enormous scale of the global oil and gas system, the past decade has seen rapid change. Even over the course of writing this book, oil and gas technology, markets, and government policy have each undergone major shifts, ones that few anticipated. No doubt such change will continue, and while the lessons of the past can help make sense of the present and anticipate the future, few are brave (or foolhardy) enough to predict the future of global energy confidently.

Nonetheless, this book is intended to provide a more nuanced understanding of the issues surrounding the industry, which you can use to make sense of new developments as they unfold. In this final chapter I'll consider some of the biggest shifts of the last few years, highlight the key issues that remain unchanged, and explore what it all might mean for the future.

VOLATILE MARKETS

Oil prices are famously volatile, and the swings in recent years rival those of any previous era of booms and busts. After peaking a little shy of \$150/bbl in 2008, then plunging below \$50 in the wake of the recession, prices clawed their way back to more than \$100 for most of 2011, 2012, and 2013. But as the global market shifted, prices plunged again from \$102 in August 2014 to a low of \$26 in February 2016.

Around that time, with oil prices hovering in the \$30/bbl range, I watched from the audience of CERAWeek as Ali Al-Naimi, then the

Saudi Arabian minister of petroleum and mineral resources, described how Saudi Arabia would not cut oil production alone to help stabilize oil prices. Instead, it seemed as though OPEC (whose most important member is Saudi Arabia) would wait out the price slump in expectation that oil production from higher-cost plays such as U.S. shale, Canadian oil sands, and other regions would falter.

Over the last several years, as prices plunged and stayed relatively low, many researchers and analysts watched expectantly to see how far shale production would fall in response. And while the number of drilling rigs and fracking crews dropped off rapidly, something funny happened: production of both oil and natural gas held up amazingly well.

Despite natural gas prices in the United States hovering near historic lows of $2 to $4 per million Btu over the past five years, natural gas production has actually *increased* by about 10 percent. This growth has been driven by the Marcellus shale, cementing Pennsylvania's place as a new natural gas powerhouse. Meanwhile, oil production from most shale regions, which grew at astounding rates from 2010 through 2015, has declined in the wake of the price crash, but far less than many had expected. By early 2017, prices were almost 50 percent below their late 2014 highs, yet oil production had declined by just 12 percent from its peak in early 2015. While production from major plays like the Eagle Ford and Bakken have declined, output from the Permian basin has actually grown despite the lower prices.[1]

Shale's resilience has altered OPEC's calculus. With prices stuck in the $30s and $40s through most of 2016, OPEC reversed course and struck a deal with several non-OPEC nations (led by Russia) to coordinate a production cut in 2017, hoping to boost prices. Global supply and demand have begun to balance, and prices have crept upward into the $50/bbl range. In the United States, rig counts have begun to rise in response to these higher prices, and oil production is likely to grow.

Despite the recent low prices, growing global demand for oil will require large-scale investment in the coming decades. In 2016, the world consumed more than 96 million barrels of petroleum products each day, and absent more aggressive international action to mitigate climate change, consumption in 2040 is projected to reach nearly 110 mb/d.[2] The recent downturn in prices has resulted in a sharp drop in spending from oil and gas companies, which invested about 25 percent less in new projects in 2015 than the previous year, with further cuts of 24 percent expected in 2016.[3]

Because U.S. shale producers can ramp up production more quickly than other projects—such as arctic or deepwater, which require lead times of five to ten years—shale production may help moderate large price swings in the future. Even so, the world may be looking at a period of oil shortages in the coming years. Because all wells become less productive over time, companies must annually produce an additional 4 to 5 mb/d just to keep production flat.[4] And if demand grows while large projects with long lead times fail to come online, shortages could be the result. As Bob McNally argues in his recent book *Crude Volatility*, global oil markets may be entering a new period of boom-bust prices, despite the success of the shale revolution.[5]

WHEN DOES A REVOLUTION END?

It can be easy to think of the shale revolution as a single breakthrough: a onetime leap forward in oilfield technology. But that's far too simple. Since around 2005, when shale gas started to make a real dent in domestic energy markets, the revolution that began in North Texas has been followed by a steady evolution playing out across the oilfields of the United States. Companies have continued to experiment and improve on the suite of technologies that led to the breakthrough in the Barnett, becoming more productive each year. Low prices provided yet another incentive to squeeze as much oil and gas as possible from each well, leading to a suite of advances ranging from the mundane to the extraordinary.

On the mundane side, companies worked hard to streamline their operations by optimizing logistics and eliminating any costs they deemed unnecessary. These efficiencies helped reduce drilling times: what used to take months began to take weeks. Companies also negotiated better prices from service firms like Halliburton or Schlumberger (service companies like these often perform the crucial steps of well casing/cementing, fracking, and other operations), who, because of slackening demand, were willing to work for less.

Companies also developed impressive technological advances. They started drilling longer laterals, driving horizontally through the shale for two miles or longer. They began mixing more sand into fracking fluids, which helped prop fractures open longer and wider. In one case, Chesapeake Energy pumped more than 25,000 tons of sand into a single well, dubbing the experiment "propageddon."[6]

Perhaps most impressively, many companies have deployed drilling rigs that don't need to be taken apart before they move to new drilling sites. Instead, these "walking rigs" can move themselves across a pad to drill multiple wells in quick succession. Resembling the towering grey walkers from Star Wars (but moving much more slowly), these behemoths use hydraulics to shift their massive frames from one spot to the next.

The result? In 2007, an average rig in the Marcellus might drill a well that produced 500 million cubic feet of natural gas per day (MMcf/d). By early 2017, the average rig drilled wells producing more than 12,000 MMcf/d, an improvement of 2,400 percent. In the Permian basin, the oil produced from an average rig grew from about 60 barrels per day in 2007 to more than 600 by 2017, a 1,000 percent increase.[7]

And there's no reason to expect that these advances won't continue, making shale more and more efficient in the coming years and decades.

TRUMP

With the election of Donald Trump in November, 2016, a number of important energy and environmental policies initiated under the Obama administration have been called into question. As I described in chapter 7, the difference between the two presidents on the topic of climate change, in particular, could hardly be more stark.

Because the Obama administration did not succeed in passing comprehensive climate legislation through Congress, the bulk of its efforts to reduce domestic greenhouse gas emissions came through executive actions. And while some executive actions are relatively straightforward to change or reverse, others require substantial time and effort before they can be eliminated entirely.

For example, the Trump administration can relatively quickly reverse an Obama-era moratorium on offering new leases for coal mining on federal land. It can also act fairly quickly to ease some rules on hydraulic fracturing on federal land developed under the previous administration. In addition, the administration has signaled an interest in revising (and, in all likelihood, lowering) the federal government's Social Cost of Carbon estimate, a calculation that underpins efforts to develop regulations on energy and climate change.[8]

But for policies that apply to the nation as a whole, the requirements of the federal rulemaking process do not make it easy or quick to reverse finalized rules. For the Clean Power Plan, the centerpiece of the Obama administration's efforts on climate change, a lengthy process will be required before the rule can officially be removed from the books and changed. In addition, courts have affirmed that the U.S. EPA *must* regulate carbon dioxide emissions under the Clean Air Act, meaning that the Trump administration will be obligated to make some effort (even if it is a toothless one) to regulate emissions. The same basic logic holds for other rules, including those for methane emissions and automobile-efficiency standards.

Nonetheless, the extent to which these rules are weakened or rolled back will certainly inform whether the shale revolution supports the fight against climate change or impedes it. As discussed in chapter 7, one of the key benefits of the shale revolution has been the availability of low-cost natural gas, which made it more politically and economically feasible to implement climate policies like the Clean Power Plan. But without those policies, most research suggests that low-cost natural gas does little to reduce long-term domestic emissions and may even lead to emissions growth. Limiting methane leaks from the oil and gas sector is another prerequisite for natural gas's ability to help mitigate climate change, and the Trump administration appears poised to weaken or roll back Obama-era methane rules.

Some have argued that continued declines in the cost of renewables such as wind and solar will make climate policy unnecessary and that the United States could have met its commitments under the 2015 Paris Agreement without such policy.[9] While this would be great news if it were true, the bulk of the research on this topic shows clearly that neither the United States nor the world as a whole is likely to reach its medium- or long-term climate goals without ambitious policies.[10]

Despite major shifts over the past several years in technology, markets, and government policy, the key issues addressed in this book will hold constant regardless of who occupies the Oval Office. Since most oil and gas regulations and tax laws are developed at the state, rather than the federal, level, a new administration in Washington will do little to alter the risks associated with methane migration, wastewater spills, induced seismicity, and more. And while federal climate policies will play a central role in the long-term trajectory of national emissions, local air-pollution risks associated with drilling, fracking, and other stages of well development will continue to be overseen and managed by state agencies.

Of course, federal policies can and will affect the industry moving forward to some extent. For example, proposed changes in federal tax policy would have a significant effect on the oil and gas industry as a whole.[11] But it is the states and local communities that play host to the industry that will continue to feel the economic booms and busts that are driven not by federal policy but by larger market forces and commodity prices.

Ultimately, the shale revolution has—for better or for worse—fundamentally changed the trajectory of the U.S. energy system. In decades to come, future generations will look back to the early 2000s and assess whether the United States seized or wasted the opportunity provided by this shift.

THE FINAL WORD

For vocal opponents of fracking, a nationwide ban is the ultimate goal. For most of the oil and gas industry and its most ardent boosters, the goal is to reduce the weight of regulation and allow the industry to create jobs, revitalize communities, and turn a profit. For the average person, it's not clear which course is best.

Hopefully, this book has made it clear that neither side of the debate has all the right answers and provided some understanding of the tangible effects of the shale revolution. These include the very real benefits, very real risks, and continuing uncertainties that will likely persist for years, if not decades, to come. Researchers will continue to look into the extent of water contamination from poorly constructed wells, the risks to human health, and the ways to prevent wastewater disposal from causing damaging earthquakes. They will also try to understand the economic effects, not just for the communities that play host to the oil and gas industry, but for the nation and the globe as a whole.

As policy makers consider the questions of whether and how fracking should continue, their decisions will create environmental, economic, and geopolitical ripples that will spread across the United States and the world. If shale development continues to play a major role in the domestic (and perhaps global) energy landscape, policy makers and the public will have to wrestle with the messy trade-offs, nuances, and uncertainties that all too often have been absent from the debate. Ultimately, any opinion about whether fracking is worth the risks will hinge on two key issues: your perspective and your tolerance for risk. If you don't stand to benefit from oil and gas development and believe that any risk to the environment or human health is unacceptable, you'll probably oppose the industry. If you stand to benefit and are willing to bear some risk, you'll probably support fracking.

Segments of industry, government, and the NGO community will continue to work to reduce the risks of pollution, but those concerns will never be eliminated entirely. At the same time, some advocates in industry and government will seek to roll back existing regulations, potentially increasing the chances of harm to the environment and human health.

Despite statewide bans in New York and Maryland and continued debate in a number of other regions, it seems likely that the broader shale revolution is here to stay. While oil and gas prices, political leadership, and oilfield technology will continue to evolve and sometimes shift dramatically, the fracking genie is out of the bottle. Looking forward, shale development (like other industries tied closely to volatile commodity prices) will come and go in unpredictable waves.

In producing regions, riding these waves of activity will be a challenge in and of itself. When I visited Williston and the surrounding area in

2013, as oil prices hovered around $100 per barrel, nearly two hundred rigs were drilling away, each boring more than a mile deep into the earth. The Wal-Mart was stripped bare, condos were sprouting up on the outskirts of town, restaurants and bars were jam-packed with free-spending Texans, and prosperity was in the air.

Three years later, with oil prices closer to $40 per barrel, there were fewer than forty drilling rigs in the Bakken. Airlines were canceling their regular flights from Houston to Williston, the Wal-Mart parking lot was quiet, and the blackjack table in my hotel bar sat empty. Many of the new condos that had gone up outside of town were now finished but virtually unoccupied, waiting for the next wave.

For places like Williston, the reemergence of the oil and gas industry will continue to bring mixed blessings. It has brought enormous wealth, but the industry's volatile nature means that the local economy will endure wide swings as long as it remains dependent on the oil industry. Like other boomtowns before it, Williston will try to solidify its gains by diversifying its economy and becoming less dependent on unpredictable oil prices. The open question is whether it will follow San Francisco, a boomtown that sprung up in the 1800s thanks to gold mining and grew into a global city, or Midland, the West Texas outpost of the oil industry, which is still cycling through each boom and bust.

And while the environmental concerns that sparked widespread opposition to fracking are unlikely to disappear anytime soon, shale development will continue to take place near people's homes and sources of drinking water. In the absence of groundbreaking new evidence about risks to the environment or public health, it's hard to imagine a nationwide, or even many statewide, bans on fracking being enacted.

Despite these many unknowns, partisans on either end of the fracking debate will continue to push their side of the story, emphasizing one set of issues while ignoring another. Moving beyond these polarized arguments will require not just a willingness to listen to the other side of the debate, but also to wrestle with a set of thorny and complex trade-offs where there are often no easy answers. With luck, a better understanding of the facts can enable this discussion, with the goal of reducing the risks, increasing the benefits, and seeking out answers to the uncertainties of the shale revolution.

NOTES

1. INTRODUCTION

1. National Cancer Institute, "Cell Phones and Cancer Risk" (2016), http://www.cancer.gov/cancertopics/causes-prevention/risk/radiation/cell-phones-fact-sheet.
2. F. A. Wilson and J. P. Stimpson, "Trends in Fatalities from Distracted Driving in the United States, 1999 to 2008," *American Journal of Public Health* 100, no. 11 (2010): 2213–2219; S. G. Klauer et al., "Distracted Driving and Risk of Road Crashes Among Novice and Experienced Drivers," *New England Journal of Medicine* 370, no. 1 (2014): 54–59; J. L. Nasar and D. Troyer, "Pedestrian Injuries Due to Mobile Phone Use in Public Places," *Accident Analysis and Prevention* 57 (2013): 91–95; D. C. Schwebel et al., "Distraction and Pedestrian Safety: How Talking on the Phone, Texting, and Listening to Music Impact Crossing the Street," *Accident Analysis and Prevention* 45 (2012): 266–271.

2. WHAT IS FRACKING?

1. D. L. Bartlett and J. B. Steele, *Howard Hughes: His Life and Madness* (New York: Norton, 2011), 29.
2. R. Gold, *The Boom: How Fracking Ignited the American Energy Revolution and Changed the World* (New York: Simon and Schuster, 2014).
3. For two examples, see B. Reed, "Nitroglycerin Explosion of 1927," *Butler County Historical*, http://butlerhistorical.org/items/show/41; and Venango County Pennsylvania Historical Society, "Oil Deaths," http://www.venango.pa-roots.com/oildeaths.html.
4. J. C. Page and J. L. Miskimins, *A Comparison of Hydraulic and Propellant Fracture Propagation in a Shale Gas Reservoir*, presented at the Canadian International Petroleum Conference 2008, Paper 2008-008.
5. C. T. Montgomery and M. B. Smith, "Hydraulic Fracturing: History of an Enduring Technology," *Journal of Petroleum Technology* 62, no. 12 (2010): 26–40.

6. Some in the industry bridle at terms like "frack job" or "fracking" because of their similarity to a lewd term and prefer to use "frac' job," "frac'ing," "fracing," or even "fraccing." Opponents of oil and gas development tend to insert the "k," perhaps in an attempt to inflame negative connotations. Throughout this book, I will use "frack" and "fracking" not because I align myself with opponents but because these terms are used much more commonly among the media and general public than those that drop the "k."
7. Throughout this book, I will refer to "shale" and "tight" formations. Fracking can be applied to any type of oil and gas reservoir, though shale and other tight formations receive the treatment in almost all cases, while conventional sources may or may not be fracked. Shale formations such as the Marcellus or Haynesville are primarily producers of natural gas. Other formations such as the Bakken are not technically "shale"; rather, they are "tight" formations that are more permeable than shale and may produce both oil and natural gas. In an effort to avoid technical discussions of various tight rocks, I will frequently use the term "shale" to refer to both types of formations.
8. For an in-depth look at how a few oil and gas companies came to develop shale economically, see the first section, entitled "Breakthrough," of G. Zuckerman, *The Frackers: The Outrageous Inside Story of the New Billionaire Wildcatters* (New York: Penguin, 2013): 17–111.
9. For overviews of the three experiments, see M. Reynolds Jr., B. G. Bray, and R. L. Mann, "Project Rulison: A Status Report," in *Society of Petroleum Engineers Eastern Regional Meeting* (Pittsburgh, Penn., 1970); D. Ward and C. Atkinson, "Project Gasbuggy: A Nuclear Fracturing Experiment," *Journal of Petroleum Technology* 18, no. 2 (1966): 139–145; and W. R. Woodruff and R. S. Guido, "Project Rio Blanco. Part I. Nuclear Operations and Chimney Reentry," Conference: 4. IAEA Panel on Peaceful Nuclear Explosives, Vienna, Austria, January 20, 1975. For a fun video showing one of the experiments, see Los Alamos National Laboratory, "Declassified U.S. Nuclear Test Film #36" (1969), https://www.youtube.com/watch?v=myXswNUQgLs.
10. M. Shellenberger, T. Nordhaus, et al., *Where the Shale Gas Revolution Came From: Government's Role in the Development of Hydraulic Fracturing in Shale* (Breakthrough Institute, 2012).
11. Zuckerman, *The Frackers*, 33–39.
12. Gold, *The Boom*, 190–192.
13. Zuckerman, *The Frackers*, 109–111.
14. I spoke with dozens of landowners around the United States about the leasing process, and this was the range I found. The $30,000-per-acre figure was secured by a local government in Louisiana that leased a large tract of land at the peak of speculation over the Haynesville shale formation.
15. In some regions, especially those that have experienced natural-resource development in prior decades, property owners may have at some point sold their surface rights and subsurface rights separately. In those cases, the owner of the surface rights may not

have any say or see any revenue if the mineral owner decides to sign a lease. However, the surface owner may experience the hassle and risks of oil and gas development occurring on their land. This has important implications for oil and gas development, not to mention the fascinating debate over whose interests win out if there is a dispute: the mineral owner or the landowner. (Under most state laws, the mineral owner wins.)

16. Interstate Oil and Gas Compact Commission, *Summary of State Statutes and Regulations* (2013).
17. Most fracking operations, especially those involving long laterals, are performed in stages. That is, operators will hydraulically fracture one section of the well (often roughly 500 feet long) at a time until they have completed the entire length of the lateral.
18. K. Fisher and N. Warpinski, "Hydraulic-Fracture-Height Growth: Real Data," *SPE Production and Operations* 27, no. 1 (2012): 8–19.
19. Production volumes based on data from a Drilling Info search for most productive wells in McKenzie County, North Dakota, from 2013 to 2015. The top producing well identified was operated by Whiting Oil and Gas, API 33-053-05634.
20. R. Sandrea and I. Sandrea, "New Well-Productivity Data Provide U.S. Shale Potential Insights," *Oil and Gas Journal* 112, no. 11 (2014).
21. TYT Politics, "Josh Fox Spotlights Community UPRISING Against Fracking" (2017), https://www.youtube.com/watch?v=Z2tBBkb8UKk.
22. S. Kelly, "Water Contamination Verdict Prompts Calls for Federal Action on Fracking," *DeSmog*, March 16, 2016, https://www.desmogblog.com/2016/03/16/dimock-water-contamination-verdict-leads-renewed-calls-federal-action-fracking.
23. K. Valentine, "Study Links Water Contamination to Fracking Operations in Texas and Pennsylvania," *ThinkProgress.org*, September 15, 2014, https://thinkprogress.org/study-links-water-contamination-to-fracking-operations-in-texas-and-pennsylvania-bfe75e731830.
24. B. D. Drollette, K. Hoelzer, et al., "Elevated Levels of Diesel Range Organic Compounds in Groundwater Near Marcellus Gas Operations Are Derived from Surface Activities," *Proceedings of the National Academy of Sciences* 112, no. 43 (2015): 13184–13189.
25. K. Brown, "New Study Finds Fracking Has Not Contaminated Drinking Water," *Energy in Depth*, October 13, 2015, https://energyindepth.org/national/new-duke-study-finds-fracking-has-not-contaminated-drinking-water/.
26. R. Bailey, "Fracking Does Not Contaminate Drinking Water, Says New Yale Study," *Frackfeed.com*, October 13, 2015, http://www.frackfeed.com/fracking-does-not-contaminate-drinking-water-says-new-yale-study/.
27. D. Levitan, "Inhofe on Fracking, Water Contamination," *Factcheck.org*, June 5, 2015, http://www.factcheck.org/2015/03/inhofe-on-fracking-water-contamination/.
28. R. Hildreth, "Leading Business Group Says Fracking Provides 'Substantial Economic Benefits' for Colorado, in Energy in Depth Mountain States," *Energy in Depth*, January 25, 2016, https://energyindepth.org/mtn-states/leading-business-group-says-fracking

-provides-substantial-economic-benefits-for-colorado/; J. Green, "Energy Secretary Ernest Moniz: Fracking Is Good for the Economy and Environment," *Energy in Depth*, August 16, 2016, https://energyindepth.org/national/energy-secretary-ernest-moniz-fracking-is-good-for-the-economy-and-environment/.

29. Gold, *The Boom*, 190–192.
30. For the definitive history of the early oil industry, see D. Yergin, *The Prize: The Epic Quest for Oil, Money, and Power* (New York: Simon and Schuster, 1990), 1–134.
31. For a review of the many times that "peak oil" fears have roiled markets, see B. C. Clayton, *Market Madness: A Century of Oil Panics, Crises, and Crashes* (Oxford: Oxford University Press, 2014).

3. DOES FRACKING CONTAMINATE WATER?

1. R. B. Jackson et al., "The Depths of Hydraulic Fracturing and Accompanying Water Use Across the United States," *Environmental Science and Technology* 49, no. 15 (2015): 8969–8976. This study also showed that about 6 percent of wells had been fracked at depths of less than 3,000 feet from the surface. I discuss some of these shallower wells later in the chapter.
2. U.S. Environmental Protection Agency, *Hydraulic Fracturing for Oil and Gas: Impacts from the Hydraulic Fracturing Water Cycle on Drinking Water Resources in the United States* (Washington, D.C., 2016).
3. N. R. Warner, R. B. Jackson, et al., "Geochemical Evidence for Possible Natural Migration of Marcellus Formation Brine to Shallow Aquifers in Pennsylvania," *Proceedings of the National Academy of Sciences* 109, no. 30 (2012): 11961–11966.
4. In 2015, the Pennsylvania Department of Environmental Protection issued a list of 248 cases where the department had determined that oil and gas activity had led to an "impact" on local water supplies. The document is available on the DEP's website, at: http://files.dep.state.pa.us/OilGas/BOGM/BOGMPortalFiles/OilGasReports/Determination_Letters/Regional_Determination_Letters.pdf.
5. U.S. Energy Information Administration, "Crude Oil and Natural Gas Exploratory and Development Wells," 2015, https://www.eia.gov/dnav/ng/NG_ENR_WELLEND_S1_A.htm.
6. R. B. Jackson, A. Vengosh, et al., "The Environmental Costs and Benefits of Fracking," *Annual Review of Environment and Resources* 39 (2014): 327–362.
7. S. G. Osborn, A. Vengosh, et al., "Methane Contamination of Drinking Water Accompanying Gas-Well Drilling and Hydraulic Fracturing," *Proceedings of the National Academy of Sciences* 108, no. 20 (2011): 8172–8176.
8. See, for example, G. T. Llewellyn, F. Dorman, et al., "Evaluating a Groundwater Supply Contamination Incident Attributed to Marcellus Shale Gas Development," *Proceedings of the National Academy of Sciences* 112, no. 20 (2015); R. D. Vidic, S. L. Brantley, et al.,

"Impact of Shale Gas Development on Regional Water Quality," *Science* 340, no. 6134 (2013); R. B. Jackson, A. Vengosh, et al., "Increased Stray Gas Abundance in a Subset of Drinking Water Wells Near Marcellus Shale Gas Extraction," *Proceedings of the National Academy of Sciences* 110, no. 28 (2013): 11250–11255.

9. V. F. Nuccio, "Coal-Bed Methane: Potential and Concerns" (U.S. Department of the Interior, U.S. Geological Survey, 2000), https://pubs.usgs.gov/fs/fs123-00/fs123-00.pdf.
10. Erie County Department of Parks, Recreation, and Forestry, "Chestnut Ridge County Park," 2015, http://www2.erie.gov/parks/index.php?q=chestnut-ridge.
11. "Fire Water: But Not the Sort You Want to Drink," *Economist* blog, 2013, http://www.economist.com/blogs/babbage/2013/06/fracking.
12. H. Li and K. H. Carlson, "Distribution and Origin of Groundwater Methane in the Wattenberg Oil and Gas Field of Northern Colorado," *Environmental Science and Technology* 48, no. 3 (2014): 1484–1491; O. A. Sherwood, J. D. Rogers, et al., "Groundwater Methane in Relation to Oil and Gas Development and Shallow Coal Seams in the Denver-Julesburg Basin of Colorado," *Proceedings of the National Academy of Sciences* 113, no. 30 (2016).
13. Llewellyn, Dorman, et al., "Evaluating a Groundwater Supply Contamination Incident."
14. U.S. Environmental Protection Agency, "Pavillion Groundwater Investigation," 2016, https://www.epa.gov/region8/pavillion.
15. D. C. DiGiulio and R. B. Jackson, "Impact to Underground Sources of Drinking Water and Domestic Wells from Production Well Stimulation and Completion Practices in the Pavillion, Wyoming, Field," *Environmental Science and Technology* 50, no. 8 (2016): 4524–4536.
16. Wyoming Department of Environmental Quality, *Pavillion Wyoming Area Domestic Water Wells Final Report and Tables* (2016), http://deq.wyoming.gov/wqd/pavillion-investigation/resources/investigation-final-report/.
17. Z. L. Hildenbrand, D. D. Carlton, et al., "Temporal Variation in Groundwater Quality in the Permian Basin of Texas, a Region of Increasing Unconventional Oil and Gas Development," *Science of the Total Environment* 562 (2016): 906–913.
18. J. Fox and the Gasland Team, *The Sky Is Pink* (2012), https://vimeo.com/44367635.
19. C. Brufatto, J. Cochran, et al., "From Mud to Cement—Building Gas Wells," *Oilfield Review* 14, no. 3 (2013).
20. President's BP Oil Spill Commission, *Deep Water: The Gulf Oil Disaster and the Future of Offshore Drilling—Report to the President* (2010), https://www.gpo.gov/fdsys/pkg/GPO-OILCOMMISSION/pdf/GPO-OILCOMMISSION.pdf.
21. A. R. Ingraffea, M. T. Wells, et al., "Assessment and Risk Analysis of Casing and Cement Impairment in Oil and Gas Wells in Pennsylvania, 2000–2012," *Proceedings of the National Academy of Sciences* 111, no. 30 (2014): 10955–10960.
22. T. Considine, R. Watson, et al., "Environmental Impacts During Marcellus Shale Gas Drilling: Causes, Impacts, and Remedies" (Buffalo: State University of New York at

Buffalo, Shale Resources and Society Institute, 2012), http://www.velaw.com/UploadedFiles/VEsite/E-comms/UBSRSI-EnvironmentalImpact.pdf.

23. G. E. King and D. E. King, "Environmental Risk Arising from Well Construction Failure: Difference Between Barrier and Well Failure, and Estimates of Failure Frequency Across Common Well Types, Locations, and Well Age," SPE 166142, in *Society of Petroleum Engineers, Annual Technical Conference and Exhibition* (New Orleans, La., 2013).
24. W. J. Calosa, B. Sadarta, and R. Ronaldi, "Well Integrity Issues in the Malacca Strait Contract Area," SPE 129083, in *Society of Petroleum Engineers, Oil and Gas India Conference and Exhibition* (Mumbai, India, 2010).
25. S. Kell, "State Oil and Gas Agency Groundwater Investigations and Their Role in Advancing Regulatory Reforms: A Two-State Review: Ohio and Texas," prepared for the Ground Water Protection Council, August 2011, https://fracfocus.org/sites/default/files/publications/state_oil__gas_agency_groundwater_investigations_optimized.pdf.
26. O. A. Sherwood, J. D. Rogers, et al., "Groundwater Methane in Relation to Oil and Gas Development."
27. Texas Groundwater Protection Committee, "Joint Groundwater Monitoring and Contamination Report—2015" (SFR-056/15, 2016), https://www.tceq.texas.gov/assets/public/comm_exec/pubs/sfr/056-15.pdf.
28. City of Beaumont Water Utilities Department, "Annual Water Quality Report" (Beaumont, Tex., 2015), http://beaumonttexas.gov/wp-content/uploads/pdf/water/water_quality_report_15.pdf.
29. S. Aronow, "Report 133: Ground-Water Resources of Chambers and Jefferson Counties, Texas" (Texas Water Development Board, 1971).
30. City of Oil City [Pennsylvania], "2014 Drinking Water Quality Report," https://www.culliganbottledwater.com/resources/water-quality-reports/docs/2014-water-quality-report-oil-city-pa.pdf.
31. A. Krupnick, H. Gordon, and S. Olmstead, "Pathways to Dialogue: What the Experts Say about the Environmental Risks of Shale Gas Development," 2013, http://www.rff.org/files/sharepoint/Documents/RFF-Rpt-PathwaystoDialogue_FullReport.pdf.
32. B. D. Drollette, K. Hoelzer, et al., "Elevated Levels of Diesel Range Organic Compounds in Groundwater Near Marcellus Gas Operations Are Derived from Surface Activities," *Proceedings of the National Academy of Sciences* 112, no. 43 (2015).
33. S. Gorman and D. Whitcomb, "Plains All American Pipeline Indicted in Santa Barbara Oil Spill," *Reuters*, May 17, 2016, http://www.reuters.com/article/us-california-oilspill-idUSKCN0Y826J.
34. D. Dekok, "Range Resources Faces Fine Over Pennsylvania Wastewater Leak," *Reuters*, August 6, 2014, http://www.reuters.com/article/2014/08/06/usa-fracking-pennsylvania-idUSL2N0QC27120140806; W. Stueck, "Leak Shuts Down Fracking-Water Storage Pond; Talisman Says Environmental Risks Are Low," *Globe and Mail*, October 30, 2013, http://www.theglobeandmail.com/news/british-columbia/leak-shuts

-fracking-water-storage-pond-talisman-says-environmental-risks-are-low/article15176909/.

35. DiGiulio and Jackson, "Impact to Underground Sources of Drinking Water."
36. D. Hopey, "Range Resources to Pay $4.15M Penalty," *Pittsburgh Post-Gazette*, September 18, 2014, http://www.post-gazette.com/local/2014/09/18/DEP-orders-Range-Resources-to-pay-4-million-fine/stories/201409180293; D. Hopey, "Shale Gas Driller Fined $1.2M for Contaminating Drinking Water in Westmoreland," *Pittsburgh Post-Gazette*, February 28, 2017, http://www.post-gazette.com/local/westmoreland/2017/02/28/WPX-Energy-Appalachia-shale-gas-company-fined-Pennsylvania-water-contamination-Westmoreland-County/stories/201702280305.
37. S. Kell, "State Oil and Gas Agency Groundwater Investigations."
38. L. A. Patterson, K. E. Konschnik, et al., "Unconventional Oil and Gas Spills: Risks, Mitigation Priorities, and State Reporting Requirements," *Environmental Science and Technology* 51, no. 5 (2017): 2563–2573.
39. S. Detrow, "Lycoming County Frack Truck Crash: 3,600 Gallons Spilled," September 27, 2012, http://stateimpact.npr.org/pennsylvania/2012/09/27/lycoming-county-frack-truck-crash-3600-gallons-spilled/; J. Cain, "3 Tanker Trucks Crash in Canton Township, Spilling Fracking Water, Diesel Fuel," April 21, 2014, http://www.wtae.com/news/tanker-trucks-crash-in-canton-township/25579248.
40. U.S. Environmental Protection Agency, "History of the UIC Program—Injection Well Time Line," 2015, http://water.epa.gov/type/groundwater/uic/history.cfm.
41. J. Cart, "Agencies Admit Failing to Protect Water Sources from Fuel Pollution," *Los Angeles Times*, March 10, 2015.
42. E. M. Gilmer, "Regulators Say Riskiest Injection Wells Have Been Shut Down," May 19, 2015, http://www.eenews.net/energywire/stories/1060018760/search?keyword=california.
43. D. M. Akob, A. C. Mumford, et al., "Wastewater Disposal from Unconventional Oil and Gas Development Degrades Stream Quality at a West Virginia Injection Facility," *Environmental Science and Technology* 50, no. 11 (2016); C. D. Kassotis, L. R. Iwanowicz, et al., "Endocrine Disrupting Activities of Surface Water Associated with a West Virginia Oil and Gas Industry Wastewater Disposal Site," *Science of the Total Environment* 557–558 (2016): 901–910; A. Lustgarten, "Injection Wells: The Poison Beneath Us," 2012, https://www.propublica.org/article/injection-wells-the-poison-beneath-us; Z. L. Hildenbrand, D. D. Carlton, et al., "A Comprehensive Analysis of Groundwater Quality in the Barnett Shale Region," *Environmental Science and Technology* 49, no. 13 (2015).
44. U.S. Environmental Protection Agency, "Class II Oil and Gas Related Injection Wells," 2016, https://www.epa.gov/uic/class-ii-oil-and-gas-related-injection-wells.
45. N. E. Lauer, J. S. Harkness, and A. Vengosh, "Brine Spills Associated with Unconventional Oil Development in North Dakota," *Environmental Science and Technology* 50, no. 10 (2016); M. Soraghan and P. King, "Drilling Mishaps Damage Water in Hundreds of Cases," *E&E News: EnergyWire*, August 8, 2016.

46. J. Hanson, "Spoiled Soil: How Salt Water Spills Are Hitting a North Dakota Farmer Hard," *KVRR* [North Dakota], 2016, http://www.kvrr.com/2016/11/24/spoiled-soil-how-salt-water-spills-are-hitting-a-north-dakota-farmer-hard/.
47. N. R. Warner, C. A. Christie, et al., "Impacts of Shale Gas Wastewater Disposal on Water Quality in Western Pennsylvania," *Environmental Science and Technology* 47, no. 20 (2013): 11849–11857; S. M. Olmstead, L. A. Muehlenbachs, et al., "Shale Gas Development Impacts on Surface Water Quality in Pennsylvania," *Proceedings of the National Academy of Sciences* 110, no. 13 (2013): 4962–4967.
48. Coverage from the *New York Times*, which featured a series of articles called "Drilling Down," spread this story widely. See, for example, I. Urbina, "Behind Veneer, Doubt on Future of Natural Gas," *New York Times*, June 26, 2011.
49. U.S. Environmental Protection Agency, "Pretreatment Standards for the Oil and Gas Extraction Point Source Category," EPA 821-F-16-001 (Office of Water, 2016), https://www.epa.gov/sites/production/files/2016-06/documents/uog-final-rule_fact-sheet_06-14-2016.pdf.
50. Indeed, there was much controversy over the U.S. EPA's use of the terms "widespread" and "systemic" in their 2015 draft and 2016 final report on the risks of hydraulic fracturing on drinking-water resources. For a brief overview of this issue, see I. Echarte, A. Krupnick, and D. Raimi, "Understanding EPA's Final Report on Hydraulic Fracturing," 2016, http://www.rff.org/blog/2016/understanding-epa-s-final-report-hydraulic-fracturing.
51. Notably, there was extensive pollution in Oil City and other early drilling regions, though the long-term impacts of this pollution appear to be fairly limited. For a description of the water-contamination issues in Flint, see J. Lapook, "Doctors Explain the Long-Term Health Effects of Flint Water Crisis," *CBS News*, January 19, 2016, http://www.cbsnews.com/news/doctors-explain-the-long-term-health-effects-of-flint-water-crisis/.

4. WILL FRACKING MAKE ME SICK?

1. DI Desktop, "Eagle%" search query, 2017. DI Desktop, an oil- and gas-industry data service, provides detailed information about oil and gas leasing, drilling, production, and more across the United States. It also allows users to search through regulatory filings for data on wells drilled into specific geological formations. Because the Eagle Ford shale sometimes goes by different spellings in regulatory filings (such as Eagle Ford or EagleFord), this search term returns all relevant results for wells drilled in the Eagle Ford.
2. Weather Channel, Inside Climate News, and Center for Public Integrity, "Fracking the Eagle Ford Shale: Big Oil and Bad Air on the Texas Prairie," *Inside Climate News*, 2014, https://insideclimatenews.org/content/fracking-eagle-ford-shale-big-oil-bad-air-texas-praire.

3. U.S. Occupational Safety and Health Administration, "Safety and Health Topics: Hydrogen Sulfide," 2017, https://www.osha.gov/SLTC/hydrogensulfide/hazards.html.
4. Operators typically use freshwater, though some have experimented with brackish water or other water sources that don't compete with uses such as irrigation and human consumption. For additional information on the composition of fracking fluids, see https://fracfocus.org/water-protection/drilling-usage.
5. E. G. Elliott, P. Trinh, et al., "Unconventional Oil and Gas Development and Risk of Childhood Leukemia: Assessing the Evidence," *Science of the Total Environment* 576 (2017): 138–147.
6. C. D. Kassotis, K. C. Klemp, et al., "Endocrine-Disrupting Activity of Hydraulic Fracturing Chemicals and Adverse Health Outcomes After Prenatal Exposure in Male Mice," *Endocrinology* 156, no. 12 (2015): 2015–1375.
7. "Fracking Could Be Giving You Massive Balls, Tiny Sperm Count," *Grist*, October 15, 2015, http://grist.org/climate-energy/fracking-could-be-giving-you-massive-balls-tiny-sperm-count/.
8. D. J. Rozell and S. J. Reaven, "Water Pollution Risk Associated with Natural Gas Extraction from the Marcellus Shale," *Risk Analysis* 32, no. 8 (2012): 1382–1393.
9. This estimate of trucks on the road comes from S. Abramzon, C. Samaras, et al., "Estimating the Consumptive Use Costs of Shale Natural Gas Extraction on Pennsylvania Roadways," *Journal of Infrastructure Systems* 20, no. 3 (2014).
10. U.S. Department of Transportation, Pipeline and Hazardous Materials Administration, "Hazmat Incident Statistics 2017," http://www.phmsa.dot.gov/hazmat/library/data-stats/incidents.
11. D. Hopey, "Utica Shale Well Blowout in Ohio Brought Under Control," *Pittsburgh Post-Gazette*, December 24, 2014, http://powersource.post-gazette.com/powersource/companies/2014/12/24/Utica-Shale-well-blowout-in-Ohio-brought-under-control/stories/201412240215.
12. "StatOil Fined $223,000 Over Ohio Fracking-Well Fire," *Columbus Dispatch*, September 16, 2015, http://www.dispatch.com/content/stories/local/2015/09/16/company-fined-over-well-mishap.html.
13. Texas Railroad Commission, "Monthly Drilling, Completion, and Plugging Summaries," 2017, http://www.rrc.state.tx.us/oil-gas/research-and-statistics/well-information/monthly-drilling-completion-and-plugging-summaries/.
14. Texas Railroad Commission, "Blowouts and Well Control Problems," 2017, http://www.rrc.state.tx.us/oil-gas/compliance-enforcement/blowouts-and-well-control-problems/.
15. A few companies have experimented with powering their generators with natural gas, which produces fewer emissions that could affect health. This point reflects the larger fact that technologies in the oilfield are constantly changing, in some cases reducing risks (for example, more pipelines have been moving water around oilfields in recent years, reducing truck traffic) and in other cases increasing them (another recent trend has been using larger volumes of sand, which increases truck traffic on the road).

16. U.S. Bureau of Labor Statistics, “Census of Fatal Occupational Injuries Charts, 1992–2015,” https://www.bls.gov/iif/oshcfoi1.htm.
17. For example, every meeting at ExxonMobil, even if it takes place in a staid office tower, begins with a safety briefing, and every oil and gas tour I’ve ever taken begins with a detailed discussion of safety.
18. States and, in some cases, local governments use regulatory tools known as “setbacks” to designate how close oil and gas development may be to homes, businesses, schools, bodies of water, and other sensitive locations. See chapter 6.
19. U.S. Environmental Protection Agency, “Health Assessment Document for Diesel Engine Exhaust,” prepared by the National Center for Environmental Assessment, Washington, D.C., for the Office of Transportation and Air Quality, 2002, EPA/600/8-90/057F, https://cfpub.epa.gov/si/si_public_file_download.cfm?p_download_id=36319.
20. U.S. National Institutes of Health, Tox Town, “Volatile Organic Compounds,” 2016, http://toxtown.nlm.nih.gov/text_version/chemicals.php?id=31.
21. U.S. Environmental Protection Agency, “2014 National Emissions Inventory (NEI) Data,” 2017, https://www.epa.gov/air-emissions-inventories/2014-national-emissions-inventory-nei-data.
22. U.S. Environmental Protection Agency, “Basic Information: Emissions from the Oil and Gas Industry, 2016,” http://www3.epa.gov/airquality/oilandgas/basic.html.
23. U.S. National Institutes of Health, Tox Town, “Volatile Organic Compounds.”
24. At least five recent studies have examined this issue, including R. F. Swarthout, R. S. Russo, et al., “Impact of Marcellus Shale Natural Gas Development in Southwest Pennsylvania on Volatile Organic Compound Emissions and Regional Air Quality,” *Environmental Science and Technology* 49, no. 5 (2015); T. Colborn, K. Schultz, et al., “An Exploratory Study of Air Quality Near Natural Gas Operations,” *Human and Ecological Risk Assessment: An International Journal* 20, no. 1 (2014): 86–105; A. Litovitz, A. Curtright, et al., “Estimation of Regional Air-Quality Damages from Marcellus Shale Natural Gas Extraction in Pennsylvania,” *Environmental Research Letters* 8, no. 1 (2013): 014017; D. T. Allen, “Atmospheric Emissions and Air Quality Impacts from Natural Gas Production and Use,” *Annual Review of Chemical and Biomolecular Engineering* 5 (2014): 55–75; L. Fleischman, J. Banks, and J. Graham, “Fossil Fumes: A Public Health Analysis of Toxic Air Pollution from the Oil and Gas Industry,” published by the Clean Air Task Force and the Alliance of Nurses for Healthy Environments, 2015, http://www.catf.us/resources/publications/files/FossilFumes.pdf; J. L. Collett Jr., J. Ham, et al., “Characterizing Emissions from Natural Gas Drilling and Well Completion Operations in Garfield County, CO,” prepared by the Colorado State University Department of Atmospheric Science, 2016, https://www.garfield-county.com/public-health/documents/CSU_GarCo_Report_Final.pdf.
25. For example, Southwestern Energy implemented green completions at its wells in Arkansas in 2010, and the Norwegian oil company Statoil implements these technologies at many of its wells in the Bakken.

26. Colorado Department of Health and Environment, "Colorado Oil and Gas Health Information and Response Program," 2015, http://www.oghir.dphe.state.co.us/.
27. Colorado Department of Health and Environment, "Assessment of Potential Public Health Effects from Oil and Gas Operations in Colorado," 2017, https://drive.google.com/file/d/0B0tmPQ67k3NVVFc1TFgleDhMMjQ/view.
28. Three often-cited examples are M. Bamberger and R. Oswald, "Unconventional Oil and Gas Extraction and Animal Health," *Environmental Science: Processes and Impacts* 16, no. 8 (2014): 1860–1865; M. Bamberger and R. Oswald, "Long-Term Impacts of Unconventional Drilling Operations on Human and Animal Health," *Journal of Environmental Science and Health* 50, no. 5 (2015): 447–459; A. Tustin, A. Hirsch, et al., "Associations Between Unconventional Natural Gas Development and Nasal and Sinus, Migraine Headache, and Fatigue Symptoms in Pennsylvania," *Environmental Health Perspectives* 125, no. 2 (2017): 189–197.
29. C. D. Kassotis, D. E. Tillitt, et al., "Estrogen and Androgen Receptor Activities of Hydraulic Fracturing Chemicals and Surface and Ground Water in a Drilling-Dense Region," *Endocrinology* 155, no. 3 (2014): 897–907.
30. University of Missouri School of Medicine, "MU Researchers Find Fracking Chemicals Disrupt Hormone Function," 2014, http://medicine.missouri.edu/news/0214.php.
31. L. M. McKenzie, R. Guo, et al., "Birth Outcomes and Maternal Residential Proximity to Natural Gas Development in Rural Colorado," *Environmental Health Perspectives* 122 (2014): 412–417.
32. See, for example, R. Lewis, "New Study Links Fracking to Birth Defects in Heavily Drilled Colorado," *Al-Jazeera America*, January 30, 2014, http://america.aljazeera.com/articles/2014/1/30/new-study-links-frackingtobirthdefectsinheavilydrilledcolorado.html.
33. Examples here include C. Warneke, F. Geiger, et al., "Volatile Organic Compound Emissions from the Oil and Natural Gas Industry in the Uintah Basin, Utah: Oil and Gas Well Pad Emissions Compared to Ambient Air Composition," *Atmospheric Chemistry and Physics* 14 (2014): 10977–10988; J. B. Gilman, B. M. Lerner, et al., "Source Signature of Volatile Organic Compounds from Oil and Natural Gas Operations in Northeastern Colorado," *Environmental Science and Technology* 47, no. 3 (2013): 1297–1305; D. Helmig, C. R. Thompson, et al., "Highly Elevated Atmospheric Levels of Volatile Organic Compounds in the Uintah Basin, Utah," *Environmental Science and Technology* 48, no. 9 (2014): 4707–4715; Z. L. Hildenbrand, P. M. Mach, et al., "Point Source Attribution of Ambient Contamination Events Near Unconventional Oil and Gas Development," *Science of the Total Environment* 573 (2016): 382–388.
34. L. M. McKenzie, R. Z. Witter, et al., "Human Health Risk Assessment of Air Emissions from Development of Unconventional Natural Gas Resources," *Science of the Total Environment* 424, no. 1 (2012): 79–87.
35. Colorado Department of Public Health and Environment, "Air Emissions Case Study Related to Oil and Gas Development in Erie, CO," 2012, http://www.colorado.gov

/airquality/tech_doc_repository.aspx?action=open&file=Erie+Air+Emissions+Case+Study+2012+-+revised+11252014.pdf.

36. L. M. McKenzie et al., "Childhood Hematologic Cancer and Residential Proximity to Oil and Gas Development," *PLoS ONE* 12, no. 2 (2017).
37. Colorado Department of Health and Environment, "Assessment of Potential Public Health Effects."
38. G. P. Macey, R. Breech, et al., "Air Concentrations of Volatile Compounds Near Oil and Gas Production: A Community-Based Exploratory Study," *Environmental Health* 13, no. 82 (2014).
39. S. L. Stacy, L. L. Brink, et al., "Perinatal Outcomes and Unconventional Natural Gas Operations in Southwest Pennsylvania." *PloS ONE* 10, no. 6 (2015): e0126425.
40. Statistical significance is a key indicator for researchers and measures how likely a particular event is to occur by chance. If there is a less than 5 percent probability of some outcome (such as low birth weight among a substantial number of newborns), and a researcher finds that the outcome has occurred, those results are said to be statistically significant. Because those results were unlikely to have occurred by chance, statistical significance suggests that the outcome was likely caused by some other factor (in this case, proximity to shale development).
41. J. A. Casey, D. A. Savitz, et al., "Unconventional Natural Gas Development and Birth Outcomes in Pennsylvania, USA," *Epidemiology* 27, no. 2 (2016).
42. T. Jemielita, G. L. Gerton, et al., "Unconventional Gas and Oil Drilling Is Associated with Increased Hospital Utilization Rates," *PLoS ONE* 10, no. 7 (2015).
43. S. G. Rasmussen, E. L. Ogburn, et al., "Association Between Unconventional Natural Gas Development in the Marcellus Shale and Asthma Exacerbations," *JAMA Internal Medicine* 176, no. 9 (2016).
44. For a couple of examples here, see T. Wheatley, "Opposition to Fracking in Maryland Is Anti-Science," *Washington Post*, March 3, 2017, https://www.washingtonpost.com/blogs/all-opinions-are-local/wp/2017/03/03/opposition-to-fracking-in-maryland-is-anti-science/; or S. Whitehead, "New Activist Report Rehashes Discredited Fracking Studies to Target School Children," *Energy in Depth*, 2016, https://energyindepth.org/national/activist-report-rehashes-discredited-fracking-studies-target-school-children/.
45. International Agency for Research on Cancer, *Air Pollution and Cancer*, ed. K. Straif, A. Cohen, and J. Samet (World Health Organization, 2013).
46. For a rich discussion of the history of coal in Europe and the United States, see B. Freese, *Coal: A Human History*, 2016 ed. (New York: Basic Books, 2013). For details on the Black Fog, see 167–168.
47. Clean Air Task Force, "Death, Disease, & Dirty Power: Mortality and Health Damage Due to Air Pollution from Power Plants," 2000, http://www.catf.us/resources/publications/files/Death_Disease_Dirty_Power.pdf.
48. Clean Air Task Force, "The Toll from Coal: An Updated Assessment of Death and Disease from America's Dirtiest Energy Sources," 2010, http://www.catf.us/resources/publications/files/The_Toll_from_Coal.pdf.

49. N. Z. Muller, R. Mendelsohn, and W. Nordhaus, "Environmental Accounting for Pollution in the United States Economy," *American Economic Review* 101 (August 2011): 1649–1675.
50. In a series of widely cited papers, Mark Z. Jacobsen and colleagues lay out arguments for a rapid transition to an energy system powered entirely by wind, water, and sun. Two papers that highlight this work are M. Z. Jacobson and M. A. Delucchi, "Providing All Global Energy with Wind, Water, and Solar Power, Part I: Technologies, Energy Resources, Quantities and Areas of Infrastructure, and Materials," *Energy Policy* 39, no. 3 (2011): 1154–1169; M. Z. Jacobson, M. A. Delucchi, et al., "100% Clean and Renewable Wind, Water, and Sunlight (WWS) All-Sector Energy Roadmaps for the 50 United States," *Energy and Environmental Science* 8, no. 7 (2015): 2093–2117.
51. A far larger array of research shows that any transition away from fossil fuels would happen more slowly, summarized by the IPCC in its most recent report and distilled succinctly in the International Energy Agency's "450" Scenario included in its annual *World Energy Outlook: Intergovernmental Panel on Climate Change, Mitigation of Climate Change*, ed. O. Edenhofer et al. (Cambridge: Cambridge University Press, 2014); and International Energy Agency, *World Energy Outlook* (Paris, 2016).

5. DOES FRACKING CAUSE EARTHQUAKES?

1. Descriptions of fracking-induced quakes for each of these locations are available through the following reports and papers: Seismological Society of America, "Fracking Confirmed as Cause of Rare 'Felt' Earthquake in Ohio," 2015, https://phys.org/news/2015-01-fracking-rare-felt-earthquake-ohio.html; UK Department of Energy and Climate Change, "Fracking UK Shale: Understanding Earthquake Risk," 2014, http://m.northyorks.gov.uk/CHttpHandler.ashx?id=32806&p=0; British Columbia Oil and Gas Commission, "Industry Bulletin 2015-32: August Seismic Event Determination," 2015, https://www.bcogc.ca/node/12951/download; A. A. Holland, "Earthquakes Triggered by Hydraulic Fracturing in South-Central Oklahoma," *Bulletin of the Seismological Society of America* 103, no. 3 (2013): 1784–1792; C. Frohlich, H. DeShon, et al., "A Historical Review of Induced Earthquakes in Texas," *Seismological Research Letters* 87, no. 4 (2016); X. Bao and D. W. Eaton, "Fault Activation by Hydraulic Fracturing in Western Canada," *Science*, November 17, 2016.
2. U.S. Environmental Protection Agency, "History of the UIC Program—Injection Well Time Line," 2015, http://water.epa.gov/type/groundwater/uic/history.cfm; D. M. Evans, "The Denver-Area Earthquakes and the Rocky Mountain Arsenal Disposal Well," *Mountain Geologist* 3, no. 1 (1966).
3. Evans, "The Denver-Area Earthquakes."

4. For a description of the Rocky Mountain Arsenal issue along with an overview of other quakes associated with energy development, see Committee on Induced Seismicity Potential in Energy Technologies, Committee on Earth Resources, et al., *Induced Seismicity Potential in Energy Technologies* (Washington, D.C.: National Academies Press, 2012).
5. K. E. Murray, "Class II Saltwater Disposal for 2009–2014 at the Annual-, State-, and County- Scales by Geologic Zones of Completion, Oklahoma," 2015, Oklahoma Geological Survey Open File Report OF5-2015, http://ogs.ou.edu/docs/openfile/OF5-2015.pdf.
6. X. C. Colazas and R. W. Strehle, "Subsidence in the Wilmington Oil Field, Long Beach, California, USA," in *Developments in Petroleum Science*, vol. 41, *Subsidence Due to Fluid Withdrawal*, ed. E. C. Donaldson et al. (Amsterdam: Elsevier, 1995), 285–335.
7. S. E. Hough and M. Page, "Potentially Induced Earthquakes During the Early Twentieth Century in the Los Angeles Basin," *Bulletin of the Seismological Society of America* 107, no. 2 (2016).
8. Frohlich, DeShon, et al., "A Historical Review of Induced Earthquakes in Texas"; S. E. Hough and M. Page, "A Century of Induced Earthquakes in Oklahoma?" *Bulletin of the Seismological Society of America* 105, no. 6 (2015).
9. Committee on Induced Seismicity Potential in Energy Technologies et al., *Induced Seismicity Potential in Energy Technologies.*
10. For example, see the following: A. McGarr, B. Bekins, et al., "Coping with Earthquakes Induced by Fluid Injection," *Science* 347, no. 6224 (2015): 830–831; W. L. Ellsworth, "Injection-Induced Earthquakes," *Science* 341, no. 6142 (2013); F. R. Walsh and M. D. Zoback, "Oklahoma's Recent Earthquakes and Saltwater Disposal," 2015, http://www.searchanddiscovery.com/pdfz/documents/2016/80516walsh/ndx_walsh.pdf.html; M. J. Hornbach, M. Jones, et al., "Ellenburger Wastewater Injection and Seismicity in North Texas," *Physics of the Earth and Planetary Interiors* 261 (2016); M. J. Hornbach, H. R. DeShon, et al., "Causal Factors for Seismicity Near Azle, Texas," *Nature Communications* 6, no. 6728 (2015).
11. States First Induced Seismicity by Injection Work Group, "Potential Injection-Induced Seismicity Associated with Oil and Gas Development, 2015: An Initiative of the Interstate Oil and Gas Compact Commission and the Groundwater Protection Council," http://www.gwpc.org/sites/default/files/finalprimerweb.pdf.
12. "4.3-Magnitude Earthquake Rattles Central Oklahoma," *Associated Press*, December 29, 2015, http://fuelfix.com/blog/2015/12/29/4-3-magnitude-earthquake-rattles-central-oklahoma/.
13. See, for example, L. Manwarren, "Large Earthquake Damages Downtown Cushing, Rattles Residents Across the State," *Oklahoma News 9*, November 6, 2016; B. Bailey, "Most Damage from 5.8-Magnitude Pawnee Earthquake Not Covered by Insurance," *The Oklahoman*, October 30, 2016.

14. See, for example, B. Whitaker, "Earthquake Capital of the Continental US: Oklahoma," *CBS 60 Minutes*, May 6, 2016; "The New Earthquake Capital of the U.S. Is . . . " *Weather Channel*, February 11, 2015, https://weather.com/science/environment/news/which-state-has-the-most-earthquakes.
15. This calculation comes through personal communications with Jeremy Boak, the director of the Oklahoma Geological Survey.
16. A. Pantsios, "Confirmed: Oklahoma Earthquakes Caused by Fracking," April 23, 2015, http://www.ecowatch.com/confirmed-oklahoma-earthquakes-caused-by-fracking-1882034344.html.
17. D. Quast, "Fear of Fracking Earthquakes Is Misplaced," *Energy in Depth*, July 27, 2013, https://energyindepth.org/california/fear-of-fracking-earthquakes-is-misplaced/.
18. Ohio Department of Natural Resources, "Preliminary Report on the Northstar 1 Class II Injection Well and the Seismic Events in the Youngstown, Ohio, Area," 2012, https://oilandgas.ohiodnr.gov/portals/oilgas/pdf/UICReport.pdf.
19. The order was signed in 2012 by Governor John Kasich, Executive Order 2012-09K.
20. Seismological Society of America, "Fracking Confirmed as Cause of Rare 'Felt' Earthquake in Ohio."
21. Kansas Department of Health and Environment, Kansas Corporation Commission, and Kansas Geological Survey, "Kansas Seismic Action Plan," September 26, 2014, amended January 21, 2015, http://www.kgs.ku.edu/PRS/Seismicity/2015/Seismic_Action_Plan.pdf.
22. Kansas Corporation Commission, Order Reducing Saltwater Injection Rates, 2015, Docket no. 15-CONS-770-CMSC.
23. Data from the U.S. Geological Survey, "Earthquake Catalogue Search," 2017, https://earthquake.usgs.gov/earthquakes/map/.
24. A detailed discussion of this issue is provided by R. Galchen, "Weather Underground: The Arrival of Manmade Earthquakes," *New Yorker*, April 13, 2015.
25. Additional details on these efforts can be found via the Oklahoma Secretary for Energy and the Environment, "What We Are Doing: Oklahoma Corporation Commission," 2016, https://earthquakes.ok.gov/what-we-are-doing/oklahoma-corporation-commission/; Oklahoma Corporation Commission, "Regional Earthquake Response Plan for Western Oklahoma," 2016, http://earthquakes.ok.gov/wp-content/uploads/2015/01/02-16-16WesternRegionalPlan.pdf.
26. This information comes through news reporting from S. Stewart, "Earthquake Damages Homes, Historic Pawnee Nation Buildings." *News Channel 4 KFOR*, September 3, 2016, http://kfor.com/2016/09/03/pawnee-earthquake-damages-homes-historic-pawnee-nation-buildings/.
27. M. D. Petersen, C. S. Mueller, et al., "One-Year Seismic-Hazard Forecast for the Central and Eastern United States from Induced and Natural Earthquakes," *Seismological Research Letters* 88, no. 3 (2017).

6. IS THERE ANY REGULATION ON FRACKING?

1. I do not know where this concern came from, and I haven't heard it repeated often among antifracking advocates. I also have not come across this concern in any of the health-impacts research discussed in chapter 4.
2. N. Richardson, M. Gottlieb, et al., "The State of State Shale Gas Regulation," Resources for the Future Report, 2013, http://www.rff.org/files/sharepoint/WorkImages/Download/RFF-Rpt-StateofStateRegs_Report.pdf.
3. See U.S. Energy Information Administration, "Sales of Fossil Fuels Produced from Federal and Indian Lands, FY 2003 through FY 2014," 2015, https://www.eia.gov/analysis/requests/federallands/pdf/eia-federallandsales.pdf.
4. See, for example, the *Gasland* film website, http://one.gaslandthemovie.com/whats-fracking/faq/regulations; or the advocacy group Clean Water Action, http://cleanwater.org/page/fracking-laws-and-loopholes.
5. U.S. Environmental Protection Agency, Safe Drinking Water Act, 2015, http://water.epa.gov/lawsregs/rulesregs/sdwa/index.cfm.
6. A fundamental piece in the literature on oil and gas regulation comes from H. Wiseman, "Untested Waters: The Rise of Hydraulic Fracturing in Oil and Gas Production and the Need to Revisit Regulation," *Fordham Environmental Law Review* 20, no. 115 (2009).
7. Ibid.
8. Ibid.
9. S. M. Olmstead, L. A. Muehlenbachs, et al., "Shale Gas Development Impacts on Surface Water Quality in Pennsylvania," *Proceedings of the National Academy of Sciences* 110, no. 13 (2013): 4962–4967.
10. U.S. Federal Register, "Oil and Natural Gas Sector: Emissions Standards for New and Modified Sources," a rule proposed by the U.S. Environmental Protection Agency on September 18, 2015.
11. N. Richardson, M. Gottlieb, et al., "The State of State Shale Gas Regulation," Resources for the Future Report, 2013, http://www.rff.org/files/sharepoint/WorkImages/Download/RFF-Rpt-StateofStateRegs_Report.pdf.
12. Because local governments handle regulations on the siting of homes and businesses, they may require setbacks from oil and gas wells. However, the state would not set this requirement.
13. Some major oil and gas fields, such as those in the Los Angeles basin, have been developed in and around major population centers. But Los Angeles is the exception rather than the rule.
14. American Petroleum Institute, "Safety Data Sheets: Petroleum Industry Practices," 2009, http://www.api.org/~/media/Files/EHS/Health_Safety/SDS_Petroleum_Industry_Practices_Feb2009.pdf.
15. See, for example, a news story on a county in New Mexico that sought to restrict oil and gas development. M. Barron, "New Mexico Federal County Frames the Issues That

Will Define Future Fights Over Local Fracking Regulation," *Baker Energy Blog*, February 19, 2015, https://www.bakerenergyblog.com/2015/02/19/new-mexico-federal-court-frames-the-issues-that-will-define-future-fights-over-local-fracking-regulation/.
16. Michael Wines, "Colorado Court Strikes Down Local Bans on Fracking," *New York Times*, May 2, 2016.
17. M. B. Baker, "Denton City Council Repeals Fracking Ban," *Fort Worth Star-Telegram*, June 16, 2015.
18. For more details on Fort Worth's oil and gas regulations, see City of Fort Worth, TX, Fort Worth City Code, Chapter 15 (2016), http://www.amlegal.com/codes/client/fort-worth_tx/.
19. One attempt comes from N. Zirogiannis, J. Alcorn, et al., "State Regulation of Unconventional Gas Development in the U.S.: An Empirical Evaluation," *Energy Research and Social Science* 11 (January 2016): 142–154.
20. Richardson et al., "The State of State Shale Gas Regulation."
21. U.S. Federal Register, "Oil and Natural Gas Sector: Emissions Standards for New and Modified Sources."

7. IS FRACKING GOOD OR BAD FOR CLIMATE CHANGE?

1. U.S. Energy Information Administration, Electricity Data Browser, "Annual Net Generation," 2017.
2. Under one set of projections, one in which the Obama administration's Clean Power Plan was implemented, coal consumption decreases substantially. Under another set, where the Clean Power Plan does not come into effect, coal consumption remains essentially flat. Natural gas grows markedly under both.
3. R. Rohde and R. Muller, "Air Pollution in China: Mapping Concentrations and Sources," *PLoS ONE* 10, no. 8 (2015).
4. Two examples would be *Energy in Depth*, http://energyindepth.org/marcellus/us-leads-in-carbon-emissions-reductions-due-to-natural-gas-despite-howarth/; and the America's Natural Gas website, http://thinkaboutit.org.
5. Detailed data on U.S. coal exports are available through the U.S. Energy Information Administration's Coal Data Browser, https://www.eia.gov/beta/coal/data/browser/.
6. This point is made by S. Afsah and K. Salcito, "U.S. Coal Exports Erode All CO_2 Savings from Shale Gas," CO_2 Scorecard Report, 2014, http://co2scorecard.org/home/researchitem/29.
7. Notably, U.S. coal exports have fallen substantially, from a peak of roughly 126 million short tons in 2012 to about 74 million in 2015. In 2008, coal exports were roughly 82 million short tons. Data on imports, exports, and production are available from the U.S. Energy Information Administration Coal Data Browser.
8. Two papers that show this clearly are R. G. Newell and D. Raimi, "Implications of Shale Gas Development for Climate Change," *Environmental Science and Technology*

48, no. 15 (2014): 8360–8368; X. Lu, J. Salovaara, and M. B. McElroy, "Implications of the Recent Reductions in Natural Gas Prices for Emissions of CO_2 from the U.S. Power Sector," *Environmental Science and Technology* 46, no. 5 (2012): 3014–3021.

9. Newell and Raimi, "Implications of Shale Gas Development for Climate Change."
10. There are many examples of this work, including F. O'Sullivan and S. Paltsev, "Shale Gas Production: Potential Versus Actual Greenhouse Gas Emissions," *Environmental Research Letters* 7, no. 4 (2012): 044030; M. Jiang, W. M. Griffin, et al., "Life Cycle Greenhouse Gas Emissions of Marcellus Shale Gas," *Environmental Research Letters* 6, no. 3 (2011): 034014; S. P. A. Brown and A. J. Krupnick, "Abundant Shale Gas Resources: Long-Term Implications for U.S. Natural Gas Markets," Resources for the Future Discussion Paper 10-41, 2010, http://www.rff.org/research/publications/abundant-shale-gas-resources-long-term-implications-us-natural-gas-markets; A. Burnham, J. Han, et al., "Life-Cycle Greenhouse Gas Emissions of Shale Gas, Natural Gas, Coal, and Petroleum," *Environmental Science and Technology* 46, no. 2 (2011): 619–627; Energy Modeling Forum, "Changing the Game? Emissions and Market Implications of New Natural Gas Supplies," EMF Report 26, Volume 1 (Stanford, Calif.: Stanford University, 2013), https://web.stanford.edu/group/emf-research/docs/emf26/Summary26.pdf; H. D. Jacoby, F. O'Sullivan, and S. Paltsev, "The Influence of Shale Gas on U.S. Energy and Environmental Policy," MIT Joint Program on the Science and Policy of Global Change Report no. 207, 2011, https://globalchange.mit.edu/publication/14563; P. Jaramillo, W. M. Griffin, and H. S. Matthews, "Comparative Life-Cycle Air Emissions of Coal, Domestic Natural Gas, LNG, and SNG for Electricity Generation," *Environmental Science and Technology* 41, no. 17 (2007): 6290–6296; T. J. Skone, J. Littlefield, and J. Marriott, "Life Cycle Greenhouse Gas Inventory of Natural Gas Extraction, Delivery, and Electricity Production," DOE/NETL-2011/1522, 2011, http://www.canadiancleanpowercoalition.com/files/4713/2630/3388/LA7%20-%20NG-GHG-LCI.pdf; T. Stephenson, J. E. Valle, and X. Riera-Palou, "Modeling the Relative GHG Emissions of Conventional and Shale Gas Production," *Environmental Science and Technology* 45, no. 24 (2011): 10757–10764; A. Venkatesh, P. Jaramillo, et al., "Uncertainty in Life Cycle Greenhouse Gas Emissions from United States Gas End-Uses and Its Effects on Policy," *Environmental Science and Technology* 45, no. 19 (2011): 8182–8189; A. Venkatesh, P. Jaramillo, et al., "Implications of Changing Natural Gas Prices in the United States Electricity Sector for SO_2, NO_x, and Life Cycle GHG Emissions," *Environmental Research Letters* 7, no. 3 (2012); C. L. Weber and C. Clavin, "Life Cycle Carbon Footprint of Shale Gas: Review of Evidence and Implications," *Environmental Science and Technology* 46, no. 11 (2012): 5688–5695.
11. Because some methane is *intentionally* vented from natural-gas wells and other infrastructure, "leakage," which implies an accidental release, can be a somewhat misleading term when referring to the entirety of oil- and gas-based methane emissions. However, because most of these emissions are unintentional, I will occasionally use the term "leakage" here.

12. For an in-depth discussion of methane's contribution to climate change, see Intergovernmental Panel on Climate Change, "IPCC Fourth Assessment Report: Climate Change 2007, Working Group I: The Physical Science Basis," 2007, https://www.ipcc.ch/publications_and_data/publications_ipcc_fourth_assessment_report_wg1_report_the_physical_science_basis.htm. For a quicker overview, see any of the recent U.S. Environmental Protection Agency "U.S. Greenhouse Gas Inventory Reports," https://www.epa.gov/ghgemissions/inventory-us-greenhouse-gas-emissions-and-sinks.
13. R. A. Alvarez, S. W. Pacala, et al., "Greater Focus Needed on Methane Leakage from Natural Gas Infrastructure," *Proceedings of the National Academy of Sciences* 109, no. 17 (2012): 6435–6440.
14. R. W. Howarth, R. Santoro, and A. Ingraffea, "Methane and the Greenhouse-Gas Footprint of Natural Gas from Shale Formations," *Climatic Change* 106, no. 4 (2011): 679–690.
15. M. Levi, "A Dispatch from the People's Climate March," September 23, 2014, http://blogs.cfr.org/levi/2014/09/23/a-dispatch-from-the-peoples-climate-march/.
16. See, for example, L. M. Cathles III, L. Brown, et al., "A Commentary on 'The Greenhouse-Gas Footprint of Natural Gas in Shale Formations' by R. W. Howarth, R. Santoro, and Anthony Ingraffea," *Climatic Change* 113, no. 2 (2012): 525–535; Newell and Raimi, "Implications of Shale Gas Development for Climate Change."
17. Mark Brownstein, EDF's Vice President for Climate and Energy described to me the details of the funding arrangements. All funds were directed to researchers at the institutions (mostly universities) that conducted the research, with industry contributions matched dollar for dollar by EDF, which does not accept donations from any corporations.
18. K. Bagley and L. Song, "EDF Recruits Sprawling Network to Fund Methane Leaks Research," *Inside Climate News*, April 8, 2015, https://insideclimatenews.org/news/07042015/edf-recruits-sprawling-network-fund-methane-leaks-research-climate-change-natural-gas-fracking.
19. See, for example, NC WARN, "COMPLAINT and Request for Investigation of Fraud, Waste, and Abuse by a High-Ranking EPA Official Leading to Severe Underreporting and Lack of Correction of Methane Venting and Leakage Throughout the U.S. Natural Gas Industry," filed on June 8, 2016, with the Office of Inspector General of the U.S. Environmental Protection Agency by NC WARN, http://www.ncwarn.org/wp-content/uploads/EPA-OIG_NCWARN_Complaint_6-8-16.pdf; or S. Horn, "Frackademia: The People and Money Behind the EDF Methane Emissions Study," *DeSmog*, September 16, 2013, https://www.desmogblog.com/2013/09/16/frackademia-people-money-behind-edf-fracking-methane-emissions-study.
20. E. Pooley, *The Climate War: True Believers, Power Brokers, and the Fight to Save the Earth* (London: Hachette, 2010), 55–61.
21. M. Kang, S. Christian, et al., "Identification and Characterization of High Methane-Emitting Abandoned Oil and Gas Wells," *Proceedings of the National Academy of Sciences* 113, no. 48 (2016): 13636–13641.

22. U.S. Department of Transportation, Pipeline and Hazardous Materials Administration, "Pipeline Mileage and Facilities," 2017, http://www.phmsa.dot.gov/pipeline/library/data-stats/pipelinemileagefacilities.
23. U.S. Energy Information Administration, "Natural Gas Processing Plants in the United States: 2010 Update," 2017, https://www.eia.gov/pub/oil_gas/natural_gas/feature_articles/2010/ngpps2009/table_1.cfm.
24. U.S. Energy Information Administration, "About U.S. Natural Gas Pipelines," 2017, https://www.eia.gov/pub/oil_gas/natural_gas/analysis_publications/ngpipeline/index.html.
25. For data, see U.S. Energy Information Administration, "Count of Electric Power Industry Power Plants, by Sector, by Predominant Energy Sources Within Plant, 2005 Through 2015," 2017, https://www.eia.gov/electricity/annual/html/epa_04_01.html; U.S. Energy Information Administration, "Number and Capacity of Petroleum Refineries," 2017, https://www.eia.gov/dnav/pet/pet_pnp_cap1_dcu_nus_a.htm. For a recent analysis of potential methane emissions from these sources, see T. N. Lavoie et al., "Assessing the Methane Emissions from Natural Gas–Fired Power Plants and Oil Refineries," *Environmental Science and Technology* 51, no. 6 (2017).
26. These data are updated annually in the U.S. EPA's annual Greenhouse Gas Inventory reports.
27. See, for example, K. Brown, "Digging Deeper Into a Limited Methane Study," *Energy in Depth Mountain States*, 2013, https://energyindepth.org/mtn-states/digging-deeper-into-a-limited-methane-study/; S. Whitehead, "Data in KIT Methane Study Contradict Researchers' Conclusions," *Energy in Depth National*, 2016, https://energyindepth.org/national/kit-methane-study-data-contradict-researchers-conclusions/.
28. Whitehead, "Data in KIT Methane Study Contradict Researchers' Conclusions."
29. See, for example, R. W. Howarth, "A Bridge to Nowhere: Methane Emissions and the Greenhouse Gas Footprint of Natural Gas," *Energy Science and Engineering* 2, no. 2 (2014): 47–60; J. Romm, "Methane Leaks Erase Climate Benefit of Fracked Gas, Countless Studies Find," February 17, 2016, https://thinkprogress.org/methane-leaks-erase-climate-benefit-of-fracked-gas-countless-studies-find-8b060b2b395d; B. McKibben, "Global Warming's Terrifying New Chemistry: Our Leaders Thought Fracking Would Save Our Climate. They Were Wrong. Very Wrong," March 23, 2016, https://www.thenation.com/article/global-warming-terrifying-new-chemistry/.
30. NC WARN, "COMPLAINT and request for investigation."
31. Some of the studies I discuss in this chapter emerged as part of EDF's coordinating effort; others were carried out independently by academics at many universities.
32. See, for example, D. T. Allen, D. W. Sullivan, et al., "Methane Emissions from Process Equipment at Natural Gas Production Sites in the United States: Liquid Unloadings," *Environmental Science and Technology* 49, no. 1 (2015); D. T. Allen, V. M. Torres, et al., "Measurements of Methane Emissions at Natural Gas Production Sites in the United States," *Proceedings of the National Academy of Sciences* 110, no. 44 (2013): 17768–17773; A. L. Mitchell, D. S. Tkacik, et al., "Measurements of Methane Emissions from Natural

Gas Gathering Facilities and Processing Plants: Measurement Results," *Environmental Science and Technology* 49, no. 5 (2015): 3219–3227; H. L. Brantley, E. D. Thoma, et al., "Assessment of Methane Emissions from Oil and Gas Production Pads Using Mobile Measurements," *Environmental Science and Technology* 48, no. 24 (2014): 14508–14515; B. K. Lamb, S. L. Edburg, et al., "Direct Measurements Show Decreasing Methane Emissions from Natural Gas Local Distribution Systems in the United States," *Environmental Science and Technology* 49, no. 8 (2015): 5161–5169.

33. K. McKain, A. Down, et al., "Methane Emissions from Natural Gas Infrastructure and Use in the Urban Region of Boston, Massachusetts," *Proceedings of the National Academy of Sciences* 112, no. 7 (2015): 1941–1946.
34. A. J. Marchese, T. L. Vaughn, et al., "Methane Emissions from United States Natural Gas Gathering and Processing," *Environmental Science and Technology* 49, no. 17 (2015).
35. D. Zavala-Araiza, D. R. Lyon, et al., "Reconciling Divergent Estimates of Oil and Gas Methane Emissions," *Proceedings of the National Academy of Sciences* 112, no. 51 (2015).
36. Four of the most widely covered were A. Karion, C. Sweeney, et al., "Methane Emissions Estimate from Airborne Measurements Over a Western United States Natural Gas Field," *Geophysical Research Letters* 40, no. 16 (2013): 4393–4397; G. Pétron, A. Karion, et al., "A New Look at Methane and Nonmethane Hydrocarbon Emissions from Oil and Natural Gas Operations in the Colorado Denver-Julesburg Basin," *Journal of Geophysical Research: Atmospheres* 119, no. 11 (2014): 6836–6852; P. O. Wennberg, W. Mui, et al., "On the Sources of Methane to the Los Angeles Atmosphere," *Environmental Science and Technology* 46, no. 17 (2012): 9282–9289; J. Peischl, A. Karion, et al., "Quantifying Atmospheric Methane Emissions from Oil and Natural Gas Production in the Bakken Shale Region of North Dakota," *Journal of Geophysical Research: Atmospheres* 121, no. 10 (2016): 6101–6111.
37. As you may have noticed by now, I like jazz. In this example, the relative sizes of the regions are in no way reflective of my equally strong appreciation for the great masters Charlie Parker or Thelonious Monk.
38. J. Peischl, T. Ryerson, et al., "Quantifying Atmospheric Methane Emissions from the Haynesville, Fayetteville, and Northeastern Marcellus Shale Gas Production Regions," *Journal of Geophysical Research: Atmospheres* 120, no. 5 (2015): 2119–2139.
39. Karion, Sweeney, et al., "Methane Emissions Estimate from Airborne Measurements Over a Western United States Natural Gas Field."
40. A. Brandt, G. Heath, et al., "Methane Leaks from North American Natural Gas Systems," *Science* 343, no. 6172 (2014): 733–735; S. M. Miller, S. C. Wofsy, et al., "Anthropogenic Emissions of Methane in the United States," *Proceedings of the National Academy of Sciences* 110, no. 50 (2013).
41. This finding is summarized in A. R. Brandt, G. A. Heath, and D. Cooley, "Methane Leaks from Natural Gas Systems Follow Extreme Distributions," *Environmental Science and Technology* 50, no. 22 (2016). Specific examples of this issue come from, among

others, R. Subramanian, L. L. Williams, et al., "Methane Emissions from Natural Gas Compressor Stations in the Transmission and Storage Sector: Measurements and Comparisons with the EPA Greenhouse Gas Reporting Program Protocol," *Environmental Science and Technology* 49, no. 5 (2015): 3252–3261; and C. Frankenberg, A. K. Thorpe, et al., "Airborne Methane Remote Measurements Reveal Heavy-Tail Flux in Four Corners Region," *Proceedings of the National Academy of Sciences* 113, no. 35 (2016).

42. D. R. Lyon, R. A. Alvarez, et al., "Aerial Surveys of Elevated Hydrocarbon Emissions from Oil and Gas Production Sites," *Environmental Science and Technology* 50, no. 9 (2016).
43. S. Conley, G. Franco, et al., "Methane Emissions from the 2015 Aliso Canyon Blowout in Los Angeles, CA," *Science*, February 25, 2016.
44. This finding is demonstrated in, among other papers, M. Omara, M. R. Sullivan, et al., "Methane Emissions from Conventional and Unconventional Natural Gas Production in the Marcellus Shale Basin," *Environmental Science and Technology* 50, no. 4 (2016); and B. K. Lamb, S. L. Edburg, et al., "Direct Measurements Show Decreasing Methane Emissions from Natural Gas Local Distribution Systems in the United States," *Environmental Science and Technology* 49, no. 8 (2015): 5161–5169.
45. For example, see D. J. Graham and S. Glaister, "The Demand for Automobile Fuel: A Survey of Elasticities," *Journal of Transport Economics and Policy* 36, no. 1 (2002): 1–25; T. Klier and J. Linn, "The Price of Gasoline and New Vehicle Fuel Economy: Evidence from Monthly Sales Data," *American Economic Journal: Economic Policy* 2, no. 3 (2010): 134–153; and L. Ulrich, "With Gas Prices Less of a Worry, Buyers Pass Hybrid Cars By," *New York Times*, May 14, 2015.
46. As depicted in figure 2.4, the US EIA's Annual Energy Outlook in 2008 foresaw natural-gas prices in the range of $6 to $7 per million British Thermal Units (mmBTU) in 2015. Actual prices were $2.62. In 2008, natural-gas prices were projected to be $6.71 per mmBTU in 2020, but 2015 projections forecast the price to be $4.59 (all dollars in real 2012 terms).
47. For an overview of the factors that contributed to the price crash, see J. Bordoff and A. Losz, "Oil Shock: Decoding the Causes and Consequences of the 2014 Oil Price Drop," *Horizons* 3 (Spring 2015): 190–206.
48. Regular reports on vehicle purchases come from the University of Michigan Transportation Research Institute, "Monthly Monitoring of Vehicle Fuel Economy and Emissions," http://www.umich.edu/~umtriswt/EDI_sales-weighted-mpg.html.
49. U.S. Energy Information Administration, "Annual Energy Outlook," 2017, https://www.eia.gov/outlooks/aeo/.
50. University of Texas at Austin Energy Institute, "The Full Cost of Electricity (FCe-)," 2016, http://energy.utexas.edu/the-full-cost-of-electricity-fce/. This study found gas and wind to be the cheapest sources, even when accounting for environmental externalities such as CO_2 emissions.

51. This example is, of course, simplified in a number of ways. A relevant note is that because wind and solar are intermittent sources and can't always be relied on to produce power when it is most needed, natural gas is often a complement, as gas-fired power plants can turn on and off relatively easily. This flexibility allows gas to fill in for renewables when the wind isn't blowing and the sun isn't shining. As a result, the addition of new wind and solar needs to be "backed up" with a complement like natural gas.
52. For additional details on state-level RPS programs, see the Database of State Incentives for Renewable Energy (DSIRE), http://www.dsireusa.org/.
53. See, for example, Newell and Raimi, "Implications of Shale Gas Development for Climate Change"; H. McJeon, J. Edmonds, et al., "Limited Impact on Decadal-Scale Climate Change from Increased Use of Natural Gas," *Nature* 514, no. 7523 (2014): 482–485; C. Hausman and R. Kellogg, "Welfare and Distributional Implications of Shale Gas," National Bureau of Economic Research, 2015, http://www.nber.org/papers/w21115.
54. These projections run through 2050, but I use the time period of 2015 to 2030 because the Clean Power Plan sets targets only through that time period. Projections for solar-electricity generation include both photovoltaic (PV) and solar thermal technologies. U.S. Energy Information Administration, "Annual Energy Outlook," 2017.
55. Intergovernmental Panel on Climate Change, "Climate Change 2014: Mitigation of Climate Change. Chapter 6: Assessing Transformation Pathways," 2014, https://www.ipcc.ch/pdf/assessment-report/ar5/wg3/ipcc_wg3_ar5_chapter6.pdf.
56. It is also possible that carbon capture and sequestration (CCS) will come to play a major role in the energy sector. CCS involves capturing CO_2 emissions from smokestacks at power plants (or other industrial facilities) and pumping it deep underground for long-term sequestration. Many forecasters estimate that CCS will be needed at a large scale to achieve the types of climate goals that have been laid out by policy makers in the 2015 Paris agreement. For details on these projections, see ibid.
57. See, for example, ibid., or the "450" Scenario as outlined by International Energy Agency, *World Energy Outlook* (Paris, 2016).
58. Intergovernmental Panel on Climate Change, *Climate Change 2014: Impacts, Adaptation, and Vulnerability*, part A: *Global and Sectoral Aspects*, contribution of Working Group II to the Fifth Assessment Report of the Intergovernmental Panel on Climate Change, ed. C. B. Field et al. (Cambridge: Cambridge University Press, 2014).
59. This point is made explicitly by Newell and Raimi, "Implications of Shale Gas Development for Climate Change"; and by Brown and Krupnick, "Abundant Shale Gas Resources."
60. Taking this argument a (very plausible) step further, we can reach an even broader conclusion: the international climate agreement reached in Paris in 2015 would not have been possible without the United States and China coming to the table with substantial commitments to limit their emissions. And without cheap natural gas, it's

unlikely the United States could have put forward a plan, centered on the Clean Power Plan, substantial enough to persuade other nations to make commitments. After the election of President Trump, the United States' participation in the Paris Agreement has become highly uncertain.

61. See, for example, A. B. Jaffe et al., "A Tale of Two Market Failures: Technology and Environmental Policy," *Ecological Economics* 54 (2005): 164–174; R. N. Stavins, "Policy Instruments for Climate Change: How Can National Governments Address a Global Problem?" *University of Chicago Legal Forum* 10 (1997): 293–330.
62. A. Epstein, "The Moral Case for Fossil Fuels," 2016, http://www.moralcaseforfossilfuels.com/.
63. International Energy Agency, *Progress Toward Sustainable Energy* (Paris, 2015).

8. WILL FRACKING MAKE THE UNITED STATES ENERGY INDEPENDENT?

1. Texas Railroad Commission, "Permian Basin Information," 2015, http://www.rrc.state.tx.us/oil-gas/major-oil-gas-formations/permian-basin/.
2. D. Yergin, *The Prize: The Epic Quest for Oil, Money, and Power* (New York: Simon and Schuster, 1990), chaps. 16–19.
3. J. R. Norvell, "Railroad Commission of Texas: Its Origin and Relation to the Oil and Gas Industry," *Texas Law Review* 40 (1961): 230.
4. For a detailed look at a variety of attempts to manage prices in oil markets, see R. McNally, *Crude Volatility: The History and Future of Boom-Bust Oil Prices*, ed. J. Bordoff (New York: Columbia University Press, 2017).
5. This is a very simplified view of how OPEC has exerted its market power over the years and does not take into account a variety of issues that make this story much more complex. For more information, see any of the following: McNally, *Crude Volatility*; D. Yergin, *The Quest: Energy, Security, and the Remaking of the Modern World* (New York: Penguin, 2011); J. M. Griffin, "OPEC Behavior: A Test of Alternative Hypotheses," *American Economic Review* 75, no. 5 (1985): 954–963; D. Gately, "A Ten-Year Retrospective: OPEC and the World Oil Market," *Journal of Economic Literature* 22, no. 3 (1984): 1100–1114.
6. For a look at how factors other than limited physical supplies contributed to these famous gasoline lines, see Yergin, *The Prize*, 599, 673–674.
7. U.S. Energy Information Administration, "Petroleum and Other Liquids: U.S. Imports of Crude Oil," http://www.eia.gov/dnav/pet/hist/LeafHandler.ashx?n=PET&s=MCRIMUS2&f=A.
8. For a wonderfully colorful sample of presidential statements on energy independence, see John Stewart's take on the *Daily Show* (2010), season 15, episode 78: http://www.cc.com/video-clips/n5dnf3/the-daily-show-with-jon-stewart-an-energy-independent-future.

9. Drilling Info, "Permian Basin production, all states," 2015. As noted earlier, Drilling Info allows users to search specific regions and formations. The data included here is based on a search for activities within Texas and New Mexico's Permian Basin.
10. U.S. Energy Information Administration, "Today in Energy: Six Formations Are Responsible for Surge in Permian Basin Crude Oil Production," 2014, http://www.eia.gov/todayinenergy/detail.cfm?id=17031.
11. See, for example, F. Ghitis, "America, the Saudi Arabia of Tomorrow," *CNN*, January 14, 2013, http://www.cnn.com/2013/01/14/opinion/ghitis-obama-energy/; and M. Handley, "Is the United States the Next Saudi Arabia?" *U.S. News and World Report*, November 12, 2012, http://www.usnews.com/news/articles/2012/11/12/is-the-united-states-the-next-saudi-arabia.
12. BP, *Statistical Review of World Energy*, 2016, http://www.bp.com/en/global/corporate/energy-economics/statistical-review-of-world-energy.html.
13. U.S. Energy Information Administration, "Petroleum and Other Liquids: Spot Prices," 2015, http://www.eia.gov/dnav/pet/pet_pri_spt_s1_m.htm.
14. National Energy Board of Canada, "Crude Oil and Petroleum Products," 2015, https://www.neb-one.gc.ca/nrg/sttstc/crdlndptrlmprdct/index-eng.html.
15. This is, of course, a very quick explanation of a very complex topic. For a quick primer on global oil markets, see the U.S. Energy Information Administration, "Oil: Crude and Petroleum Products Explained, Oil Prices and Outlook," 2017, https://www.eia.gov/energyexplained/index.cfm?page=oil_prices; for a deeper dive, see B. Clayton, *Market Madness* (Oxford: Oxford University Press, 2016).
16. J. Bordoff and A. Losz, "Oil Shock: Decoding the Causes and Consequences of the 2014 Oil Price Drop," *Horizons* 3 (Spring 2015): 190–206.
17. The topic of energy subsidies is explored in detail in International Energy Agency, "Energy Subsidies," 2015, http://www.worldenergyoutlook.org/resources/energy subsidies/; M. Kojima and D. Koplow, "Fossil Fuel Subsidies," Policy Research Working Paper 7220 (World Bank, 2015), http://documents.worldbank.org/curated/en/961661467990086330/pdf/WPS7220.pdf; B. Larsen and A. Shah, "World Fossil Fuel Subsidies and Global Carbon Emissions," WPS 1002 (World Bank, 1992), http://documents.worldbank.org/curated/en/332991468739452719/pdf/multi-page.pdf; J.-M. Burniaux and J. Chateau, "Mitigation Potential of Removing Fossil Fuel Subsidies," OECD Economics Department Working Paper 853, 2011, http://www.oecd-ilibrary.org/docserver/download/5kgdx1jr2plp-en.pdf?; and V. J. Schwanitz, F. Piontek, et al., "Long-Term Climate Policy Implications of Phasing Out Fossil Fuel Subsidies," *Energy Policy* 67 (2014): 882–894.
18. Historical energy data from BP, *Statistical Review of World Energy*.
19. U.S. Energy Information Administration, "Natural Gas: U.S. Imports by Country," 2015, http://www.eia.gov/dnav/ng/ng_move_impc_s1_m.htm; U.S. Energy Information Administration, "Natural Gas Consumption by End Use," 2015, http://www.eia.gov/dnav/ng/ng_cons_sum_dcu_nus_m.htm.

20. This issue is explored in detail by G. Zuckerman, *The Frackers: The Outrageous Inside Story of the New Billionaire Wildcatters* (New York: Penguin, 2013), 208–214, 278–282, 336–339; and by Yergin, *The Quest*, chap. 15.
21. See, for example, R. Lenzer, "Natural Gas Equals Energy Independence and Economic Rejuvenation," *Forbes*, June 20, 2012; T. Mullaney, "U.S. Energy Independence Is No Longer Just a Pipe Dream," *USA Today*, May 15, 2012; V. Rao, *Shale Gas: The Promise and the Peril* (Durham, N.C.: RTI, 2012).
22. This dynamic is shown in virtually every projection for energy consumption over the coming decades. One prime example would be the International Energy Agency's *World Energy Outlook*, 2016.
23. Imports from other nations, particularly Nigeria, have fallen more rapidly than imports from the Persian Gulf. This is largely because Nigerian imports consist mostly of light, sweet crude, the same type of oil that is produced from domestic shale formations. As a result, it is relatively easy for refineries to switch between oil from Nigerian imports to U.S. shale but more complex to switch from Persian Gulf imports (which tend to be heavier and contain more sulfur) to domestic shale oil.
24. M. L. O'Sullivan, "Energy Independence Alone Won't Boost U.S. Power," *Bloomberg View*, February 14, 2013, http://www.bloombergview.com/articles/2013-02-14/-energy-independence-alone-won-t-boost-u-s-power.
25. This quote is elided because it included an assertion that the United States was a net energy exporter in 2015, which is incorrect. While net energy imports have fallen dramatically in the past decade, the United States remained a large net importer of energy in 2015, primarily because of crude-oil imports. Despite this inaccuracy, the point made by this quote remains valid. Vice Admiral Jon Miller, "Our New Maritime Strategy and the Fifth Fleet," 2015, http://navylive.dodlive.mil/2015/03/15/our-new-maritime-strategy-and-the-fifth-fleet/.
26. See, for example, G. Eric, K. Sarica, and W. E. Tyner, "Analysis of Impacts of Alternative Policies Aimed at Increasing U.S. Energy Independence and Reducing GHG Emissions," *Transport Policy* 37 (2015): 121–133; V. J. Karplus, S. Paltsev, et al., "Should a Vehicle Fuel Economy Standard Be Combined with an Economy-Wide Greenhouse Gas Emissions Constraint? Implications for Energy and Climate Policy in the United States," *Energy Economics* 36 (2013): 322–333; B. Johansson, "Security Aspects of Future Renewable Energy Systems—A Short Overview," *Energy* 61 (2013): 598–605; and International Energy Agency, *Energy Technology Perspectives* (IEA, 2015).

9. IS FRACKING GOOD FOR THE ECONOMY?

1. M. Reynolds Jr., B. G. Bray, and R.L. Mann, "Project Rulison: A Status Report," in *Society of Petroleum Engineers Eastern Regional Meeting* (Pittsburgh, Penn., 1970).
2. H. A. Tewes, "Survey of Gas Quality Results from Three Gas-Well-Stimulation Experiments by Nuclear Explosions," Lawrence Livermore Laboratory, UCRL-52656, 1979.

3. U.S. Department of Energy, "Rulison, Colorado, Site," Office of Legacy Management, 2016, https://www.lm.doe.gov/rulison/Sites.aspx.
4. Oil shale is different from the shale gas and tight oil discussed in this book. These rocks are essentially immature or "uncooked" versions of the types of rocks that will, with hundreds of millions of years of pressure and heat, produce oil and gas. The technical potential of these rocks to produce oil is enormous, but—despite decades of effort—oil and gas companies have not been able to make the economics of oil shale work in the United States.
5. J. T. Bartis, T. LaTourrette, et al., "Oil Shale Development in the United States: Prospects and Policy Issues," RAND Corporation, 2005, http://www.rand.org/content/dam/rand/pubs/monographs/2005/RAND_MG414.pdf.
6. BBC Research and Consulting, "City of Rifle: A Case Study of Community Renewal, Growth, and Change in Northwest Colorado," prepared for the City of Rifle by BBC Research and Consulting, Northwest Colorado Socioeconomic Analysis and Forecasts, 2008, http://www.rifleco.org/DocumentCenter/View/1113.
7. The U.S. Census Bureau uses a coding system called the North American Industry Classification System to sort and label different types of industries. Here, I refer to NAICS number 211, "oil and gas extraction." Other oilfield workers can be found in NAICS category 213, "support activities for mining." However, I do not include this category because it incorporates economic activity associated not just with oil and gas but also with the mining of coal, metals, and other minerals.
8. U.S. Department of Commerce, Bureau of Economic Analysis, "Gross Domestic Product (GDP) by Industry Data, Value Added," 2015, http://www.bea.gov/industry/gdpbyind_data.htm.
9. Here's a healthy sampling of the studies examining national and regional economic effects: Perryman Group, "A Decade of Drilling: The Impact of the Barnett Shale on Business Activity in the Surrounding Region and Texas: An Assessment of the First Decade of Extensive Development," prepared for the Fort Worth Chamber of Commerce, 2011; B. Lewandowski and R. Wobbekind, "Assessment of the Oil and Gas Industry: 2012 Industry Economic and Fiscal Contributions in Colorado," Business Research Division, Leeds School of Business, University of Colorado–Boulder, 2013; D. Bangsund and N. M. Hodur, "Williston Basin 2012: Projections of Future Employment and Population, North Dakota Summary," North Dakota State University Agribusiness and Applied Economics Report no. 704, 2013, http://ageconsearch.umn.edu/bitstream/142589/2/AAE704%20web%20.pdf; D. Bangsund and N. M. Hodur, "Petroleum Industry's Economic Contribution to North Dakota in 2011," North Dakota State University Center for Agribusiness and Applied Economics Report no. 710, 2013, https://www.westernenergyalliance.org/wp-content/uploads/2013/08/Petroleum-Industrys-Economic-Contribution-to-North-Dakota-in-2011-March-2013.pdf; T. J. Considine, R. Watson, and S. Blumsack, "The Economic Impacts of the Pennsylvania Marcellus Shale Natural Gas Play: An Update," Pennsylvania State University, Department of Energy and Mineral Engineering, 2010, http://marcelluscoalition.org/wp

-content/uploads/2010/05/PA-Marcellus-Updated-Economic-Impacts-5.24.10.3.pdf; J. Oyakawa, H. Eid, et al., "Eagle Ford Shale: Economic Impact for Counties with Active Drilling," University of Texas at San Antonio, Center for Community and Business Research, 2012, http://ccbr.iedtexas.org/download/eagle-ford-shale-economic-impact-for-counties-with-active-drilling-executive-summary/; T. Tunstall, "Eagle Ford and the State of Texas," in *Economics of Unconventional Shale Gas Development*, ed. W. E. Hefley and Y. Wang (Switzerland: Springer International Publishing, 2015), 121–148; T. Tunstall and J. Oyakawa, "Economic Impact of Oil and Gas Activities in the West Texas Energy Consortium Study Region," University of Texas at San Antonio, Center for Community and Business Research, 2013, http://iedtexas.org/wp-content/uploads/2015/02/WestTexasEnergyConsort_Dec2013.pdf; T. Tunstall and J. Oyakawa, "Economic Impact of the Eagle Ford Shale," University of Texas at San Antonio Center for Community and Business Research, Institute for Economic Development, 2014, http://www.eagleford.org/images/eagle-ford-economic-impact-2014-UTSA-Appendix.pdf; University of Arkansas Center for Business and Economic Research, "Describing the Economic Impact of the Oil and Gas Industry in Arkansas," produced for AIPRO by the University of Arkansas Sam M. Walton College of Business, Fayetteville, Ark., 2009, http://cber.uark.edu/files/The_Economic_Impact_of_the_Oil_and_Gas_Industry_in_Arkansas.pdf; University of Arkansas Center for Business and Economic Research, "Revisiting the Economic Impact of the Natural Gas Activity in the Fayetteville Shale: 2008–2012," University of Arkansas Sam M. Walton College of Business, Fayetteville, Ark., 2012, http://cber.uark.edu/files/Revisiting_the_Economic_Impact_of_the_Fayetteville_Shale.pdf; Booz Allen Hamilton, "Wyoming Oil and Gas Economic Contribution Study," Casper, Wy., 2008, https://www.westernenergyalliance.org/sites/default/files/Wyoming%20Oil%20and%20Gas%20Economic%20Impact%20Study%20-%202008.PDF%3B.pdf; K. Hardy and T. W. Kelsey, "The Shale Gas Economy in the Northeast Pennsylvania Counties," in *Economics of Unconventional Shale Gas Development*, ed. W. E. Hefley and Y. Wang (Switzerland: Springer International Publishing, 2015), 71–91; C. Hausman and R. Kellogg, "Welfare and Distributional Implications of Shale Gas," National Bureau of Economic Research, 2015, http://www.nber.org/papers/w21115; T. M. Komarek, "Labor Market Dynamics and the Unconventional Natural Gas Boom: Evidence from the Marcellus Region," *Resource and Energy Economics* 45 (2016).

10. U.S. Chamber of Commerce, "What If America's Energy Renaissance Never Actually Happened?" U.S. Chamber of Commerce Institute for 21st Century Energy, Energy Accountability Series, 2016, http://www.energyxxi.org/what-if%E2%80%A6america%E2%80%99s-energy-renaissance-never-actually-happened.
11. Perryman Group, "The Economic and Fiscal Contribution of the Barnett Shale," 2014, https://www.perrymangroup.com/wp-content/uploads/Perryman-Barnett-Shale-Impact-8-7-2014.pdf.
12. Tunstall and Oyakawa, "Economic Impact of the Eagle Ford Shale."

13. See, for example, M. P. Mills, "Where the Jobs Are: Small Businesses Unleash America's Energy Employment Boom," Manhattan Institute's Power & Growth Initiative no. 4, 2014, https://www.manhattan-institute.org/html/where-jobs-are-small-businesses-unleash-energy-employment-boom-6029.html; or "Energy from Shale: Fueling Job Creation," 2016, http://www.energyfromshale.org/articles/fueling-job-creation.
14. Again, there are many studies on this issue with mixed findings. For a sampling, see H. Allcott and D. Keniston, "Dutch Disease or Agglomeration? The Local Economic Effects of Natural Resource Booms in Modern America," National Bureau of Economic Research Working Paper 20508, 2014, http://www.nber.org/papers/w20508; J. Dubé and M. Polèse, "Resource Curse and Regional Development: Does Dutch Disease Apply to Local Economies? Evidence from Canada," *Growth and Change* 46, no. 1 (2014); G. D. Jacobsen and D. P. Parker, "The Economic Aftermath of Resource Booms: Evidence from Boomtowns in the American West," *Economic Journal* 126, no. 593 (2014); K. Kuralbayeva and R. Stefanski, "Windfalls, Structural Transformation, and Specialization," *Journal of International Economics* 90, no. 2 (2013): 273–301; J. Brown, "Production of Natural Gas from Shale in Local Economies: A Resource Blessing or Curse?" Federal Reserve Bank of Kansas City, 2014, https://www.kansascityfed.org/publicat/econrev/pdf/14q1Brown.pdf; J. Haggerty, P. H. Gude, et al., "Long-Term Effects of Income Specialization in Oil and Gas Extraction: The U.S. West, 1980–2011," *Energy Economics* 45 (2014); D. G. Freeman, "The 'Resource Curse' and Regional U.S. Development," *Applied Economics Letters* 16, no. 5 (2009): 527–530; A. James and D. Aadland, "The Curse of Natural Resources: An Empirical Investigation of U.S. Counties," *Resource and Energy Economics* 33, no. 2 (2011): 440–453; J. G. Weber, "A Decade of Natural Gas Development: The Makings of a Resource Curse?" *Resource and Energy Economics* 37 (2014): 168–183.
15. For a couple of critiques, see S. Christopherson, "The False Promise of Fracking and Local Jobs," 2015, http://public-accountability.org/2015/08/frackademia/.
16. A. W. Bartik, J. Currie, et al., "The Local Economic and Welfare Consequences of Hydraulic Fracturing," National Bureau of Economic Research Working Paper 23060, 2017, http://www.nber.org/papers/w23060.
17. These estimates come from the U.S. Bureau of Labor Statistics through a downloadable tool developed by Headwaters Economics called EPS-HDT: http://headwatersec-onomics.org/tools/eps-hdt.
18. The opportunities and risks of developing economic-growth strategies with the energy sector are described in S. Carley and S. Lawrence, *Energy-Based Economic Development: How Clean Energy Can Drive Development and Stimulate Economic Growth* (London: Springer-Verlag, 2014).
19. J. Feyrer, E. T. Mansur, and B. Sacerdote, "Geographic Dispersion of Economic Shocks: Evidence from the Fracking Revolution," *American Economic Review* 107 (2017): 1313–1334.

20. J. P. Brown, T. Fitzgerald, and J. G. Weber, "Capturing Rents from Natural Resource Abundance: Private Royalties from U.S. Onshore Oil & Gas Production," *Resource and Energy Economics* 46 (2016): 23–38.
21. T. W. Kelsey and K. Hardy, "Marcellus Shale and the Commonwealth of Pennsylvania," in *Economics of Unconventional Shale Gas Development*, ed. W. E. Hefley and Y. Wang (Switzerland: Springer International Publishing, 2015), 93–120; K. Hardy and T. W. Kelsey, "Local Income Related to Marcellus Shale Activity in Pennsylvania," *Community Development* 46, no. 4 (2015).
22. For descriptions of this type of growth in North Dakota and Pennsylvania, see Center for Social Research at North Dakota State University, "2012 North Dakota Statewide Housing Needs Assessment: Housing Forecast," North Dakota Housing Finance Agency, 2012, http://www.visionwestnd.com/documents/NDSHNA_HousingFore cast_Final_000.pdf; Lycoming County, "The Impacts of the Marcellus Shale Industry on Housing in Lycoming County," Williamsport, Penn., 2012, http://www.lyco.org /Portals/1/PlanningCommunityDevelopment/Documents/MarcellusShaleHousing Study.pdf.
23. See, for example, H. Brueck, "Rents Nearing $4,500 in San Francisco Drive More Tech Employees to These Cities," *Fortune*, April 5, 2016, http://fortune.com/2016/04/05/san -francisco-rent-drives-tech-employees-out/.
24. A. Newcomb, "Costliest Place for Renters Has Yellowstone River View," *ABC News*, February 17, 2014, http://abcnews.go.com/US/life-williston-north-dakota-expensive -place-rent-apartment/story?id=22549192.
25. As noted in the previous chapter, a helpful source here is G. Zuckerman, *The Frackers: The Outrageous Inside Story of the New Billionaire Wildcatters* (New York: Penguin, 2013).
26. U.S. Energy Information Administration, "Annual Energy Review," table 3.5, "Consumer Expenditure Estimates for Energy by Source," 2012, https://www.eia.gov/totale nergy/data/annual/.
27. Also noted in the previous chapter, a good source here is J. Bordoff and A. Losz, "Oil Shock: Decoding the Causes and Consequences of the 2014 Oil Price Drop," *Horizons* 3 (Spring 2015): 190–206.
28. This interesting finding comes in the form of a working paper from C. Baumeister and L. Kilian, "Lower Oil Prices and the U.S. Economy: Is This Time Different?" Brookings Papers on Economic Activity, Conference Draft, September 15–16, 2016, https:// www.brookings.edu/bpea-articles/lower-oil-prices-and-the-u-s-economy-is-this -time-different/.
29. Congressional Budget Office, "The Economic and Budgetary Effects of Producing Oil and Natural Gas from Shale," 2014, https://www.cbo.gov/publication/49815.
30. As of this writing, the CBO has not updated this study in light of lower oil and natural-gas prices. With the ensuing downturn in oil prices, these budgetary benefits may have declined. On the other hand, projections for future domestic production since 2014, when the CBO report was written, have generally increased. For example, the

U.S. EIA's 2017 estimate for domestic oil production in the year 2040 was roughly 38 percent higher than its estimate in 2014.

31. Data on federal revenues from energy production on federal land is available through the Department of Interior's Office of Natural Resources Revenue statistical information website: http://statistics.onrr.gov/.
32. Government Accountability Office, "Oil and Gas Royalties: The Federal System for Collecting Oil and Gas Revenues Needs Comprehensive Reassessment," 2008, http://www.gao.gov/products/GAO-08-691.
33. This topic is explored by G. E. Metcalf, "The Impact of Removing Tax Preferences for U.S. Oil and Gas Production," Council on Foreign Relations, 2016, https://www.cfr.org/report/impact-removing-tax-preferences-us-oil-and-gas-production; J. E. Aldy, "Money for Nothing: The Case for Eliminating U.S. Fossil Fuel Subsidies," *Resources Magazine* April (2014); G. E. Metcalf, *U.S. Energy Tax Policy* (Cambridge: Cambridge University Press, 2010); and M. Allaire and S. Brown, "Eliminating Subsidies for Fossil Fuel Production: Implications for U.S. Oil and Natural Gas Markets," Resources for the Future Issue Brief 09-10, 2009, http://www.rff.org/files/sharepoint/WorkImages/Download/RFF-IB-09-10.pdf.
34. Texas Comptroller of Public Accounts, *Texas Comprehensive Annual Financial Report* (Austin, Tex., 2015), https://comptroller.texas.gov/transparency/reports/comprehensive-annual-financial/2015/.
35. North Dakota Tax Commissioner, *2014 State and Local Taxes: An Overview and Comparative Guide* [colloquially known as the "Redbook"] (Bismarck, N.D., 2014), https://www.nd.gov/tax/tax-resources/major-publications--info.
36. Because Alaska's government is so heavily reliant on oil-industry revenue, recent declines in production coupled with low oil prices have forced the government to cut the annual "dividend" by roughly 50 percent.
37. Texas University Lands, "The Permanent University Fund (PUF)," 2017, http://www.utlands.utsystem.edu/puf.aspx.
38. D. Raimi and R. G. Newell, "U.S. State and Local Oil and Gas Revenues," Resources for the Future Discussion Paper 16-50, 2016, http://www.rff.org/files/document/file/RFF-DP-16-50.pdf.
39. D. Raimi and R. G. Newell, "Shale Public Finance: Local Government Revenues and Costs Associated with Oil and Gas Development," National Bureau of Economic Research Working Paper no. 21542, 2015, http://www.nber.org/papers/w21542; D. Raimi and R. G. Newell, "Net Fiscal Impacts of Oil and Gas Development for Local Governments in Eight States," National Bureau of Economic Research Working Paper no. 21615, May 2016, http://www.nber.org/papers/w21615.
40. Raimi and Newell, "U.S. State and Local Oil and Gas Revenues."
41. Raimi and Newell, "Shale Public Finance"; Raimi and Newell, "Net Fiscal Impacts of Oil and Gas Development."
42. D. Raimi and R. G. Newell, "Colorado's Piceance Basin: Variation in the Local Public Finance Effects of Oil and Gas Development," Duke University Energy Initiative

Working Paper, May 2016, http://www.rff.org/research/publications/colorados-piceance-basin-variation-local-public-finance-effects-oil-and-gas.

43. Oil, of course, is considered a liquid. Natural-gas "liquids," or NGLs, typically refer to hydrocarbons that are separated from a stream of raw (gaseous) natural gas and condensed into their liquid state. For example, methane (the primary component of natural gas) is transported and sold as a gas; ethane, propane, butane, and pentane (also considered to be part of the natural gas family) are typically considered NGLs.
44. Haynes and Boone LLP, "Oil Patch Bankruptcy Monitor," updated September 9, 2016, http://www.haynesboone.com/publications/energy-bankruptcy-monitors-and-surveys.
45. Employment data are from the U.S. Bureau of Labor Statistics, "Current Employment Statistics Reports," 2016, https://www.bls.gov/ces/.
46. Baumeister and Kilian, "Lower Oil Prices and the U.S. Economy."
47. The issue of increased automation is explored in C. Krauss, "Texas Oil Fields Rebound from Price Lull, but Jobs Are Left Behind," *New York Times*, February 19, 2017.
48. E. U. Cascio and A. Narayan, "Who Needs a Fracking Education? The Educational Response to Low-Skill Biased Technological Change," National Bureau of Economics Working Paper no. 21359, 2015, http://www.nber.org/papers/w21359; D. S. Rickman et al., "Is Shale Development Drilling Holes in the Human Capital Pipeline?" *Energy Economics* 62 (2016).
49. For an interesting portrait of some of these workers, see the film *The Overnighters*, which follows several economic migrants looking for opportunity in the booming Bakken in the early 2010s.
50. J. D. Sachs and A. M. Warner, "The Curse of Natural Resources," *European Economic Review* 45, no. 4 (2001): 827–838.

10. WILL FRACKING SPREAD AROUND THE WORLD?

1. M. Shellenberger et al., "Where the Shale Gas Revolution Came From: Government's Role in the Development of Hydraulic Fracturing in Shale," Breakthrough Institute, 2012, https://thebreakthrough.org/blog/Where_the_Shale_Gas_Revolution_Came_From.pdf.
2. G. Zuckerman, *The Frackers: The Outrageous Inside Story of the New Billionaire Wildcatters* (New York: Penguin, 2013).
3. T. Boersma, M. Vandendriessche, and A. Leber, "Shale Gas in Algeria: No Quick Fix," Brookings Institution Policy Brief 15-01, 2015, https://www.brookings.edu/research/shale-gas-in-algeria-no-quick-fix/.
4. D. Fig and S. Scholvin, "Fracking the Karoo: The Barriers to Shale Gas Extraction in South Africa Based on Experiences from Europe and the U.S.," in *A New Scramble for Africa*, ed. S. Scholvin (London: Routledge, 2015), chap. 8.

5. U.S. Energy Information Administration, "Shale Gas Production Drives World Natural Gas Production Growth," 2016, http://www.eia.gov/todayinenergy/detail.cfm?id=27512.
6. BP, "Statistical Review of World Energy," 2016, http://www.bp.com/en/global/corporate/energy-economics/statistical-review-of-world-energy.html.
7. P. Andrews-Speed and C. Len, "China Coalbed Methane: Slow Start and Still Work in Progress," National University of Singapore, Energy Studies Institute Policy Brief 4, 2014, http://esi.nus.edu.sg/publications/esi-publications/publication/2015/12/18/china-coalbed-methane-slow-start-and-still-work-in-progress.
8. U.S. Energy Information Administration, "Shale Gas Development in China Aided by Government Investment and Decreasing Well Cost," 2015, http://www.eia.gov/todayinenergy/detail.cfm?id=23152.
9. A. Guo, "China's Shale Gas Reserves Jump Fivefold as Output Lags Target," *Bloomberg*, April 6, 2016, https://www.bloomberg.com/news/articles/2016-04-06/china-s-shale-gas-reserves-jump-fivefold-as-output-lags-target.
10. U.S. Energy Information Administration, "Shale Gas Production Drives World Natural Gas Production Growth."
11. For two articles that review these issues, see L. Tian et al., "Stimulating Shale Gas Development in China: A Comparison with the U.S. Experience," *Energy Policy* 75 (2014): 109–116; Z. Wan, T. Huang, and B. Craig, "Barriers to the Development of China's Shale Gas Industry," *Journal of Cleaner Production* 84 (2014): 818–823.
12. The "450" here refers to the atmospheric concentration of carbon dioxide. More specifically, researchers have estimated that concentrations of 450 parts per million of CO_2 likely correspond with an average global temperature increase of 2 degrees Celsius above preindustrial levels by the end of the twenty-first century. This "2 degrees" target is a commonly discussed goal in the context of international climate negotiations such as the 2015 Paris Agreement.
13. http://beta.fortune.com/global500/.
14. This quote comes from roughly 6:45 into an interview at Columbia University's Center on Global Energy Policy with Fu Chengyu, the former chair of Sinopec, on China's energy future. Columbia Energy Exchange podcast, July 8, 2016, http://energypolicy.columbia.edu/fu-chengyu-former-chair-sinopec-chinas-energy-future-71816.
15. B. Freese, *Coal: A Human History*, 2016 ed. (New York: Basic Books, 2003).
16. D. Yergin, *The Prize: The Epic Quest for Oil, Money, and Power* (New York: Simon and Schuster, 1990).
17. For an excellent overview of the EU Emissions Trading System, see A. D. Ellerman, C. Marcantonini, and A. Zaklan, "The European Union Emissions Trading System: Ten Years and Counting," *Review of Environmental Economics and Policy* 10, no. 1 (2016): 89–107.
18. The question here is whether renewable-energy mandates or subsidies reduce emissions at the lowest possible cost compared with economy-wide market-based systems such as cap-and-trade or carbon taxes. Most economists would argue that the latter

would be more economically efficient, though there may be other reasons, such as political support, to favor renewable-energy subsidies. For an overview of this issue, see P. Lehmann and E. Gawel, "Why Should Support Schemes for Renewable Electricity Complement the EU Emissions Trading Scheme?" *Energy Policy* 52 (2013): 597–607; and R. Stavins, "Will Europe Scrap Its Renewables Target? That Would Be Good News for the Economy and for the Environment," *An Economic View of the Environment*, January 18, 2014, http://www.robertstavinsblog.org/2014/01/18/will-europe-scrap-its-renewables-target-that-would-be-good-news-for-the-economy-and-for-the-environment/.

19. UK Department of Energy and Climate Change, "International Domestic Energy Prices," released May 26, 2016, https://www.gov.uk/government/statistical-data-sets/international-domestic-energy-prices.
20. UK Department of Energy and Climate Change, "Energy Trends and Prices," October 26, 2017, https://www.gov.uk/government/statistics/announcements/energy-trends-and-prices-august-2017.
21. UK Department of Business, "EIS Guidance on Fracking: Developing Shale Gas in the UK," 2017, https://www.gov.uk/government/publications/about-shale-gas-and-hydraulic-fracturing-fracking/developing-shale-oil-and-gas-in-the-uk.
22. "Balcombe Fracking Protest Eviction Order Granted," *BBC News*, November 11, 2013, http://www.bbc.com/news/uk-england-sussex-24904960.
23. D. Gayle, "Antifracking Protestors Evicted from Cheshire Camp," *Guardian*, January 12, 2016, https://www.theguardian.com/environment/2016/jan/12/anti-fracking-protesters-dig-in-as-eviction-battle-starts-in-upton-cheshire.
24. UK Committee on Climate Change, "Onshore Petroleum: The Compatibility of UK Onshore Petroleum with Meeting the UK's Carbon Budget," 2016, presented to Parliament pursuant to Section 49 of the Infrastructure Act 2015, https://www.theccc.org.uk/publication/onshore-petroleum-the-compatibility-of-uk-onshore-petroleum-with-meeting-carbon-budgets/.
25. These figures are drawn from four studies: IOD, "Infrastructure for Business: Getting Shale Gas Working," 2013, https://www.igasplc.com/media/3067/iod-getting-shale-gas-working-main-report.pdf; National Grid, "Future Energy Scenarios: Oil, Gas, and Electricity Transmission," 2016, http://www2.nationalgrid.com/uk/industry-information/future-of-energy/future-energy-scenarios/; Poyry and Cambridge Econometrics, "Macroeconomic Effects of European Shale Gas Production," a report to the International Association of Oil and Gas Producers (OGP), 2013, http://www.poyry.co.uk/sites/poyry.co.uk/files/public_report_ogp__v5_0.pdf; J. Broderick et al., "Shale Gas: An Updated Assessment of Environmental and Climate Change Impacts," University of Manchester, Tyndall Centre for Climate Change Research, 2011, http://www.mace.manchester.ac.uk/media/eps/schoolofmechanicalaerospaceandcivilengineering/newsandevents/news/research/pdfs2011/shale-gas-threat-report.pdf.
26. UK Committee on Climate Change, "Onshore Petroleum."

27. P. Gonzalez and J. Carroll, "Exxon CEO Says Argentina Shale Investment May Exceed $10B," 2016, http://www.rigzone.com/news/oil_gas/a/144881/Exxon_CEO_Says_Argentina_Shale_Investment_May_Exceed_10B.
28. M. Blake, "How Hillary Clinton's State Department Sold Fracking to the World," *Mother Jones*, September/October 2014, http://www.motherjones.com/environment/2014/09/hillary-clinton-fracking-shale-state-department-chevron.
29. A. Higgins, "Russian Money Suspected Behind Fracking Protests," *New York Times*, November 30, 2014, https://www.nytimes.com/2014/12/01/world/russian-money-suspected-behind-fracking-protests.html; U.S. Office of the Director of National Intelligence, "Background to 'Assessing Russian Activities and Intentions in Recent U.S. Elections': The Analytic Process and Cyber Incident Attribution," 2017, https://www.dni.gov/files/documents/ICA_2017_01.pdf.
30. H. Clinton, "Secretary Clinton on Energy Diplomacy in the 21st Century," U.S. Department of State Office of the Spokesperson, 2012, https://www.youtube.com/watch?v=OYbYsxw4_-k.
31. U.S. Department of State, "Global Shale Gas Initiative," 2010, http://www.state.gov/r/pa/prs/ps/2010/08/146161.htm.

11. DO PEOPLE LIVING NEAR FRACKING LOVE IT OR HATE IT?

1. I last spoke with Martin in mid-2016, and he was still living in Zahl. Because of the slowdown brought on by low prices, the community has had a chance to "catch up," and things were less hectic than at our first meeting in 2013.
2. Pennsylvania Department of Environmental Protection, "Consent Order and Agreement," 2009, http://files.dep.state.pa.us/OilGas/OilGasLandingPageFiles/FinalCO&A121510.pdf.
3. S. Phillips, "Dimock: A Town Divided," *StateImpact Pennsylvania*, March 28, 2012, https://stateimpact.npr.org/pennsylvania/2012/03/28/dimock-a-town-divided/.
4. A. Maykuth, "Jury Awards $4.24 to Two Families in Dimock Tainted-Water Case," *Philadelphia Inquirer*, March 10, 2016.
5. B. Doucette, "Plant Closure Brings Gloom to Coalgate," November 10, 2002, http://newsok.com/article/2814480.
6. J. Stevenson, "Gun Manufacturer Brings Jobs to Coalgate, OK," June 1, 2010, http://www.kxii.com/home/headlines/95382519.html.
7. N. E. Lauer, J. S. Harkness, and A. Vengosh, "Brine Spills Associated with Unconventional Oil Development in North Dakota," *Environmental Science and Technology* 50, no. 10 (2016).
8. FBI Uniform Crime Reporting System, https://www.ucrdatatool.gov/.
9. The Aspen Institute's dialogues are conducted on a not-for-attribution basis, allowing participants to speak freely and on behalf of themselves rather than on behalf of the

organizations they work for. I asked Rick for permission to share this story, and he graciously agreed.

10. Parts of each of these states have seen oil and gas development for decades, while in other regions, the shale revolution has meant a new scale and intensity of industry activity.
11. Apache Corporation, "Apache Corporation Discovers Significant New Resource Play in Southern Delaware Basin," Houston, Tex., 2016, http://investor.apachecorp.com/releasedetail.cfm?ReleaseID=988060.
12. R. C. Stedman et al., "Marcellus Shale Gas Development and New Boomtown Research: Views of New York and Pennsylvania Residents," *Environmental Practice* 14, no. 4 (2012): 382–393.
13. J. Kriesky et al., "Differing Opinions About Natural Gas Drilling in Two Adjacent Counties with Different Levels of Drilling Activity," *Energy Policy* 58 (2013): 228–236.
14. H. Boudet et al., "The Effect of Industry Activities on Public Support for 'Fracking,'" *Environmental Politics* 25, no. 4 (2016).
15. N. H. Paydar et al., "Fee Disbursements and the Local Acceptance of Unconventional Gas Development: Insights from Pennsylvania," *Energy Research and Social Science* 20, October (2016): 31–44.
16. D. Mukherjee and M. A. Rahman, "To Drill or Not to Drill? An Econometric Analysis of U.S. Public Opinion," *Energy Policy* 91 (2016): 341–351; B. H. Bishop, "Focusing Events and Public Opinion: Evidence from the Deepwater Horizon Disaster," *Political Behavior* 36 (2014): 1–22.
17. C. E. Clarke et al., "How Geographic Distance and Political Ideology Interact to Influence Public Perception of Unconventional Oil/Natural Gas Development," *Energy Policy* 97 (2016): 301–309.

12. WHAT'S NEXT?

1. Production data from U.S. EIA Drilling Productivity Report, April 2017. Price data from U.S. EIA, West Texas Intermediate spot price (nominal dollars per barrel). https://www.eia.gov/dnav/pet/pet_pri_spt_s1_d.htm.
2. 2016 estimates come from the International Energy Agency, "Oil Market Report" (Paris, March 2017), https://www.iea.org/oilmarketreport/omrpublic/; The 2040 projection references the New Policies Scenario from the International Energy Agency, "Global Energy Outlook" (Paris, 2016), http://www.iea.org/newsroom/news/2016/november/world-energy-outlook-2016.html.
3. International Energy Agency, "World Energy Investment" (Paris, 2016), https://www.iea.org/newsroom/news/2016/september/world-energy-investment-2016.html.
4. Estimates of global decline rates vary. However, multiple studies have recently estimated global average decline rates of 4 to 5 percent of global production, which trans-

lates roughly to 4 to 5 million barrels per day of decline each year. See, for example, "Analysis: Decline Rates, Spending Crunch Fuels Fresh Oil Industry Concerns," *Platts*, 2016, http://www.platts.com/latest-news/oil/london/analysis-decline-rates-spending-crunch-fuels-26507001; and "CERA-IHS Unveil Worldwide Oil Field Decline Rates at 4.5% Per Year," *Alexander's Gas & Oil Connection*, 2008, http://www.gasandoil.com/news/features/aa2d38c4348b8eec83f34e9c61e6efbe.

5. B. McNally, *Crude Volatility* (New York: Columbia University Press, 2016).
6. J. Carroll and D. Wethe, "Chesapeake Energy Declares 'Propageddon' with Record Frack," *Bloomberg Markets*, October 20, 2016, https://www.bloomberg.com/news/articles/2016-10-20/chesapeake-declares-propageddon-with-record-frack-in-louisiana.
7. U.S. EIA Drilling Productivity Report, April 2017, https://www.eia.gov/petroleum/drilling/.
8. For an overview on the development of the Social Cost of Carbon, see M. Greenstone, E. Kopits, and A. Wolverton, "Developing a Social Cost of Carbon for U.S. Regulatory Analysis: A Methodology and Interpretation," *Review of Environmental Economics and Policy* 7, no. 1 (2013).
9. See, for example, M. Bloomberg, "Climate Progress, with or Without Trump," *New York Times*, March 31, 2017.
10. See modeling scenarios in the Intergovernmental Panel on Climate Change, "Climate Change 2014: Impacts, Adaptation, and Vulnerability. Part A: Global and Sectoral Aspects," contribution of Working Group II to the *Fifth Assessment Report of the Intergovernmental Panel on Climate Change*, ed. C. B. Field et al. (Cambridge: Cambridge University Press, 2014); or the "450" Scenario as outlined by the International Energy Agency, "World Energy Outlook."
11. For an examination of one proposed change to federal tax policy, see J. Bordoff, "What Would the U.S. Border Tax Adjustment Mean for Energy?" *Petroleum Economist*, March 14, 2017.

INDEX